Thumbs, Toes, and Tears

Thumbs, Toes, and Tears

And Other Traits
That Make Us Human

Chip Walter

Walker & Company
New York

Copyright © 2006 by Chip Walter

Published by Walker Publishing Company, Inc., New York
Distributed to the trade by Holtzbrinck Publishers

All papers used by Walker & Company are natural, recyclable products made from
wood grown in well-managed forests. The manufacturing processes conform
to the environmental regulations of the country of
origin.

Library of Congress Cataloging-in-Publication Data has been applied for.

ISBN-10: 0-8027-1527-3
ISBN-13: 978-0-8027-1527-2

Visit Walker & Company's Web site at www.walkerbooks.com

First U.S. edition 2006

1 3 5 7 9 10 8 6 4 2

Typeset by Westchester Book Group
Printed in the United States of America by Quebecor World Fairfield

This book is dedicated to my parents,

Bill and Rosemary,

who never once told me to stop asking "why."

C. W.

Contents

Prologue

W E ARE—ALL OF US—freaks of nature. We don't generally see ourselves this way, of course. After all, being human, what could be more ordinary than a human being? But it turns out that our personal (and biased) impressions that we are unremarkable simply don't stand up against the plain, objective facts. The way we walk, for example, teetering on long, paired stilts of articulated bone, is unique among mammals, and as preposterous in its way as elephant trunks and platypus feet. We also communicate by tossing oddly intricate noises at one another, which somehow carry complex packages of feeling, thought, and information. We share and understand these sounds as if they were scents drifting on the wind, and our minds special noses that sniff the fragrance of their meaning. Using them we are able to change one another's minds, even bring one another to tears. We also invent, to the point of being dangerous, incessantly bending the things, living and otherwise, around us to our own ends. Because of this habit, we have, for better or worse, created national economies, erected the pyramids of Giza and Chichén Itzá, fashioned exquisite art, sculpture, and music, invented the steam engine, moon rockets, the digital computer, stealth bombers, and "weaponized" diseases. Nothing on the planet seems to escape our urge to remake it. These days we are even tailoring genes to remake ourselves.

This book is about how we became the strange creatures we are, and why we do these peculiarly human things. It wonders what makes us cry, why we fall in love, invent, deceive, laugh uproariously with close friends, and kiss the ones we care about. It asks what evolutionary twists and turns set in motion events that made the symphonies of Mozart, the insights and art of Leonardo, the drama, humor, and poetry of Shakespeare possible, not to mention bad soap operas, Hollywood movies, and London musicals. It speculates on why chimpanzees, despite sharing so much of our DNA, do not reflect upon the meaning of life, or if they do, why they haven't so far shared their insights. In the end it wonders how you became you and how our species became, of all the species it could have become, the thoroughly unprecedented one it is.

Human beings are insatiably curious, especially when it comes to the subject of ourselves. This is not a new insight. Philosophers, poets, theologians, and scientists from Plato to Darwin, St. Augustine to Freud have already penned volumes about our humanness that bow endless rows of the sturdiest library shelves. You might ask, if these thinkers have fallen gasping to the mat trying to wrestle these questions into submission, why this book should have any better luck. The simple answer is that today we have far more solid information to work with.

During the past decade enormous strides have been made in two broad scientific fields: genetics and neurobiology. Advances in genetics are helping us gain insights into the way all living things evolve and develop. Each of us has come to exist in the unique form we do because of the combinations of genes that our parents passed along. You are, to a large degree, the person you are because of the messages these genes sent, and continue to send, to the ten thousand trillion cells that have assembled just so to form you.[1] Hardly a day goes by without some news about a remarkable discovery that further illuminates the molecular machinery of the DNA that makes life possible.

The other field is brain research. Being a human being (as opposed to a wasp or a fruit fly), all of your behaviors and actions are not dictated by your genes alone. Your brain holds many of the secrets that make humans human. Genes may be outrageously complicated, but the human brain makes our genetic code look like the crayon drawings of a four-year-old.[2] Though it weighs a mere three pounds, it consists of a hundred billion neurons, each of which is connected in a thousand different ways to the other neurons around it. This means that every waking moment your brain is linked along a hundred trillion separate paths, trafficking in thought and insight, processing great streams of sensory input, running the complex plumbing of your body, generating (but not always resolving) all of your colliding and conflicting emotions, conscious and unconscious. These connections, by one estimate, make your possible states of mind during the course of your life greater than all of the electrons and protons in the universe.[3] Given the immensity of this number, you are never likely to think all of the thoughts you are actually capable of thinking, nor feel every possible feeling. Nevertheless, each shining day we give it a try.

Over the past decade scientists have been developing ways to scan and reveal in increasingly refined detail how our brains are constructed and operate. They are far from resolving its mysteries, but we know much more today about its behavior than we did even a short time ago. Positron Emission Tomography (PET) scanning and fMRI (Functional Magnetic Resonance Imaging) are revealing "movies" of our thoughts, or more precisely the flow of chemicals in the brain as we think and feel. Today we have a far better understanding of how language, laughter, and thought play themselves out in the brain than we did as recently as the turn of the twenty-first century. Right now the resolution of these movies is cellular, but they will soon reveal the brain at a molecular level, making the reading of minds much more than a parlor trick.

Scientists also keep nibbling away at the mysterious edges of paleoanthropology, psychology, physiology, sociology, and computer science, to mention only a handful, shedding light bit by bit on the special brand of behaviors we call human. In other words, we remain largely unknown to ourselves, but we are making impressive progress.

. . .

How did we become *human* beings? All living things are unique. The forces that drive evolution make them so, honing each down to the razor edge of itself, providing it with a handful of qualities that distinguish it as the only animal of its kind. The elephant has its trunk. Bombardier beetles manufacture and precisely shoot boiling hot toxic chemicals from their tails. Peregrine falcons have wings that propel them unerringly through the air at seventy miles an hour to their catch. These traits define these creatures and determine the way they act. But what unique traits shape and define us?

I have whittled it down to six, each unique to our kind: our big toes, our thumbs, our uniquely shaped pharynx and throat, laughter, tears, and kissing. How, you may ask, can something as common as a big toe, as silly as laughter, or as obvious as a thumb, possibly have anything to do with our ability to invent writing, express joy, fall in love, or bring forth the genius of ancestral China? What could they have to say about rockets and radio, symphonies, computer chips, tragedy, or the spellbinding art of the Sistine Chapel? Just this.

The origin of all these human accomplishments can be traced to these traits, each of which marks a fork in the evolutionary road where we went one way and the rest of the animal kingdom went the other, opening small passageways on the peculiar geography of the human heart and mind, marking trailheads that lead to the tangled outback of what makes us tick.

Take the knobby big toes we find at the ends of our feet. If they hadn't begun to straighten and strengthen more than five million years ago our ancestors would never have been able to stand upright, and their front feet would never have been freed to become hands. And if our hands had not been freed we would not have evolved the opposed and specialized thumbs we have, which made the first tools possible.

Both our toes and thumbs are linked to the third trait—our unusual throats and the uniquely shaped pharynx inside, which enables us to make more precise sounds than any animal. Standing up straightened and elongated our throats so that our voice box dropped. In time that made speech possible, but we also needed a brain that could generate the complex mental constructions that language and speech demand. Because toolmaking required a brain that could manipulate objects, it supplied the neural foundations for logic, syntax, and grammar so that eventually it could not only take objects and arrange them in an orderly manner, it also could conceive ideas for our pharynx to transform into the sound symbols we call words and organize them so they made sense as well.

A mind capable of language is also a self-aware mind. Consciousness melded our old primal drives with our newly evolved intelligence in entirely unexpected ways that even language couldn't successfully articulate. This explains the origins of laughter, kissing, and crying. Though we can glimpse their origins in the hoots, calls, and ancient behaviors of our primate cousins, no other species carries these particular arrows in the quivers they use to communicate.

. . .

Some may argue that we cannot possibly be reduced to six of anything. And some may argue that these traits are not unique to us. Kangaroos stand upright, after all. And dogs whimper and whine. And don't chimpanzees pucker and smack their lips? Yes, but kangaroos hop, they don't stride; dogs do not cry tears of sorrow or joy or pride. In fact, they don't cry any tears at all. No other animal does, not even elephants, contrary to some apocryphal stories. And while chimps can be trained to kiss, they do not naturally climb, during their adolescence, into the backseats of Chevrolets, or anything else for that matter, to neck.

The larger point is that the extraordinary abilities and behaviors that define us—for better or worse—as a species come from somewhere, and if we keep asking, "where, how, why . . ." enough, we arrive at their roots. The investigation of one illuminates the other, and together, in the peculiar arithmetic of evolution, they eventually add up to the strange, astonishing, and perplexing creatures we are. Maybe the point isn't so much to pin ourselves beneath the unforgiving glass of a microscope to arrive at definitive and irrefutable answers. We are far too complex a race to be reduced to the sum of so many split hairs. Maybe the important thing is to simply keep asking interesting questions and follow where the answers take us. As it turns out, they take us to some remarkable and fascinating places.

—Chip Walter
Pittsburgh, 2006

I
Toes

Chapter 1

The Curious Tale of the Hallux Magnus

The upright posture of man was the start of his fateful development.
—Sigmund Freud

THE STORY OF THE HUMAN RACE is exceedingly long and crooked, and it begins in the grasslands of an immense and mysterious continent.

Standing on the Serengeti plains of East Africa, you can't help but feel small. Mostly this is because, in the face of so much of the world all around, you become acutely aware of your mortal insignificance. There is no end to the grasslands, scattered shrub, and baobab trees that sweep out to the horizon. Everything—mountains, trees, gorges, clouds—shrinks. Smaller objects, such as lions and wildebeests and zebra, disappear altogether in the heat and the piddling ability of your eyes to pick out details this small.

The vast plains feel like another world, yet this is home because it was in a place something like this that our kind stumbled into its upright position and began its long journey into the present. And maybe that is why it also feels timeless, as if it has always been there and always will be.

But Africa's savannas have not always been here. Six million years ago Africa was a much more tropical place. In fact, the whole world was. Rain forests rose to latitudes as far north as London. Areas that today are dry grasslands, even deserts, were lush, tropical jungles where all varieties of ape species lived an Edenlike existence. It would be another million years before the creatures that would eventually spawn the human race split off from those that later led to chimpanzees and gorillas.[1] But in those days the primate lines had not yet diverged and the environment was warm and sheltered; the food plentiful; and predators, relatively speaking, few.

If the behavior of today's gorillas and chimpanzees is any indication, troops of primates, perhaps thirty or forty at a time, lived and roamed the forest, knuckle-walking on all fours for short jaunts, and swinging from tree to tree if they wanted to cover more territory with greater speed. Long-armed and bowlegged, their bodies were built for the jungle. Their feet, like their hands, were designed for gripping branches, with four long fingerlike toes and a fifth inner toe that acted like a thumb as they grasped and swayed gracefully from one branch to the next.

Few fossils of the tree-dwellers who preceded us remain. The moisture and bacteria that make jungles junglelike are not generally kind to bones left behind. But the evidence suggests that like today's gorillas and chimps, they very likely fashioned no tools and communicated with a limited repertoire of calls, hoots, and grunts.* Occasionally a little chest-banging might have been in order, or some fang-bearing, to clarify a particular point. Whatever the ways they communicated, these apes epitomized animal intelligence. At the time, they were the smartest primates on the planet, even if from our vantage point—had we been there to watch—their day-to-day life would have seemed monumentally simple and completely bereft of civilization. There were no torches or fire. The nights would have been utterly black and unlit except for the ever-changing moon and the brilliant stardust of the Milky Way tossed by the Big Bang across the black blanket of the sky. The world then was devoid of humanity. But humanity was coming.

. . .

Earth is a testy and capricious planet. Continents shift, mountains rise, ocean currents slip directions from north to south or east to west, the land splits and explodes and collides. This geological restlessness is one of the reasons life on Earth is so wild and rampant. Under the pressures of evolution, life remakes itself and finds a place in the fresh niches the planet

* Chimps and gorillas sometimes use grass and sticks and rocks as tools, but they do not create tools from scratch.

continually creates. Adaptation begets new species on the one hand, and on the other, wipes out those that can't adjust.

Changeable as Earth is, six million years ago it seems to have been particularly restive. Ice caps were forming over the Antarctic. Global temperatures were dropping, and so were sea levels, which blocked oceans' equatorial currents, first in the eastern Mediterranean, then at Gibraltar, and finally at the Isthmus of Panama, which rose out of a descending ocean between the nascent North and South Americas. The Mediterranean was draining and then refilling and draining again, depending on the whims of ocean currents and Earth's wandering landmasses. When it emptied, mountains of salt more than a mile high sometimes formed in its basin, only to be leveled again in a subsequent aeon.

As all of this unfolded on the western flank of the planet, landmasses in the Indian Ocean were riding northward on the backs of enormous tectonic plates. The Indian subcontinent continued to ram into southern Asia as it had done for thirty-five million years, and by then had made its way twelve hundred miles into the body of the continent, shoving up the Tibetan Plateau and the Himalayas ahead of it like plowed snow.

This shift was also pulling what scientists call "the Indonesian valve" farther north. This valve consisted of islands and landmasses that controlled the flow of billions of tons of water from the North Pacific into the Indian Ocean, and acted like a lock-and-dam assembly. But as sea levels descended and the valve's large islands edged northward, cooler water from the North Pacific began to stream south, refrigerating the African coast. Slowly the continent's climate changed as chillier winds swept off the Indian Ocean.[2] Jungles backtracked and the Sahara began to form where once there had been forests and grasslands.

These retreats weren't made rapidly or in neat lines. Scientists have recently found fossilized remnants and isotopes from six-million-year-old vegetation that reveal that regions of Ethiopia south of Egypt were wetter and more forested than previously thought. These were not tropical jungles, but they were not wide-open savannas either. Forests continued to cluster along rivers and cling to mountain valleys, and many remained on the plains for thousands of years before Serengeti-like grasslands began to nudge them completely aside.

As the plains of East Africa grew cooler and drier, the enormous continent was also breaking apart. Africa and Asia had begun parting ways, creating the Red Sea; the Gulf of Aden; and farther inland, the Great Rift Valley,

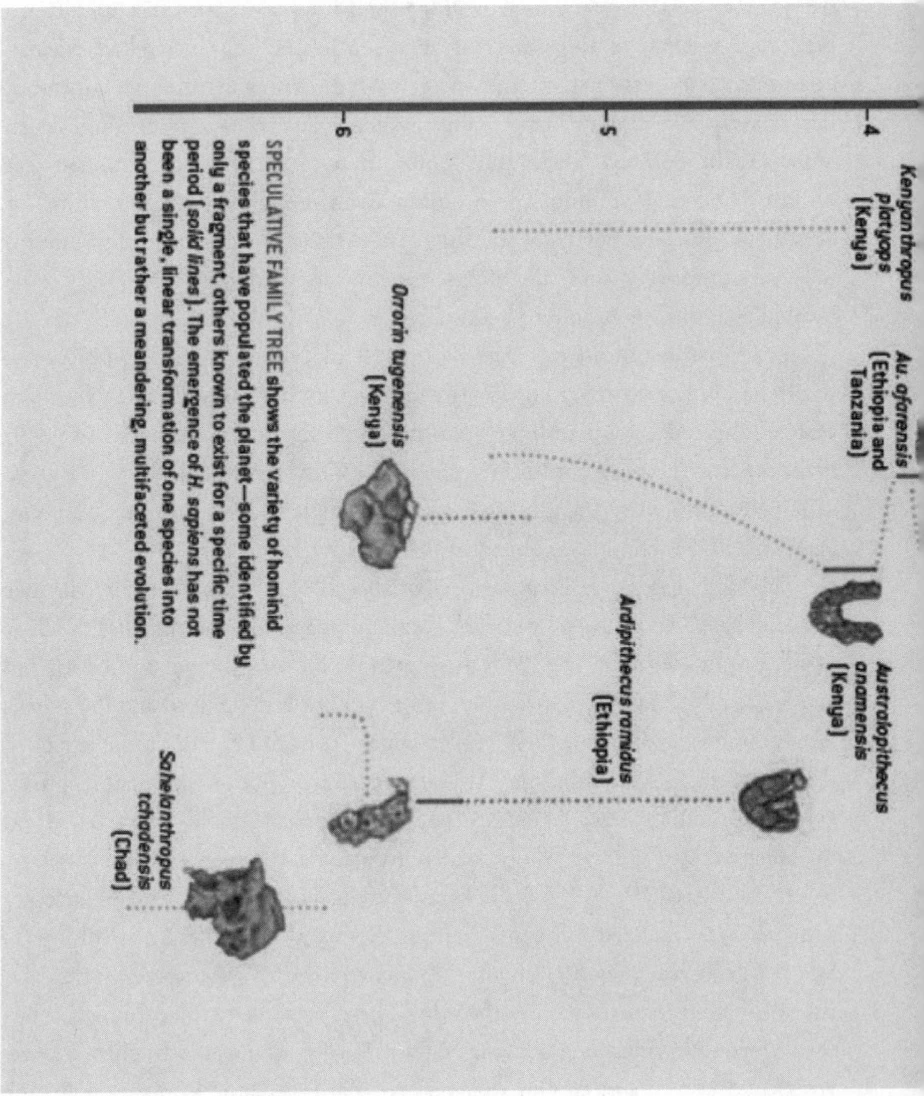
4
5
6
Kenyanthropus platyops (Kenya)
Au. afarensis (Ethiopia and Tanzania)
Australopithecus anamensis (Kenya)
Ardipithecus ramidus (Ethiopia)
Orrorin tugenensis (Kenya)
Sahelanthropus tchadensis (Chad)
SPECULATIVE FAMILY TREE shows the variety of hominid species that have populated the planet—some identified by only a fragment, others known to exist for a specific time period (solid lines). The emergence of H. sapiens has not been a single, linear transformation of one species into another but rather a meandering, multifaceted evolution.

A timeline of hominid evolution from *Sahelanthropus tchadensis* more than six million years ago to *Homo sapiens*. (Patricia Wynne, © *Scientific American*.)

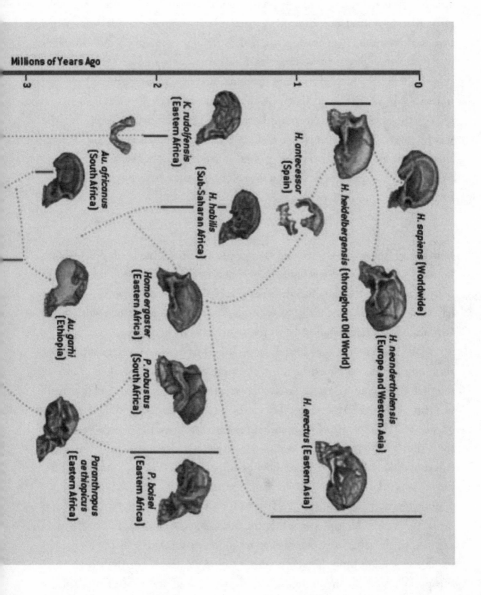

a geological scar that runs more than three thousand miles from Syria to southern Mozambique. Colossal volcanoes rose and oozed and exploded, spreading lava, smoke, and ash across thousands of square miles. Kilimanjaro, Africa's tallest mountain, is a particularly dramatic result—hardened layers of lava piled over three and a half miles high, topped by snow and glaciers even though the mountain sits almost precisely on the equator. Elsewhere the earth cracked like a great bone, and as the rift formed it opened massive valleys with walls that rose thousands of feet.

For the apes that had been hunkered down peacefully in this part of the world for so long, it must have been a confounding time as the forests shrunk, the air changed, and the ground shifted, literally, beneath their feet. What they couldn't know was how profoundly these changes would transform the apes themselves.

. . .

Because of these shifts, we are descended from a branch of the primate family that became geographically orphaned. Though anthropologists verbally brawl over the details,[3] the creatures from which we evolved seem to have been a breed of ape that was trapped on the drier, eastern side of the Rift Valley when it began to form five million years ago or so.

While our progenitors found themselves forced to cluster around what shrinking forests remained in the east, some primates followed the familiar safety of the jungles retreating to southern and central Africa beyond the western rim of the Rift Mountains. They evolved into today's three subspecies of gorillas and two species of chimpanzees—*Pan troglodytes,* the chimps you see at the circus, and bonobos *(Pan paniscus),* very likely our closest relatives. They still make their homes in these jungles, though they may not survive much longer as we deforest their habitats and hunt them down for food or sale.

But given the circumstances five million years ago, it would have been perfectly reasonable to expect that the other apes, those caught on the eastern side of the rift, were the ones marked for extinction. Yet they managed to survive, and in time they even split into several species of their own. Anthropologists didn't understand this, even a few decades ago. They had assumed that our kind had marched cleanly from a single line of protohumans into the present, like new models of cars or computers, each subsequent version an improvement over the previous one. We know better now, though the picture is far from perfectly clear.

Some species, for example, may have evolved in Chad,[4] much farther west than anthropologists once thought. A handful may even have met and interbred as they wandered the drying, broken landscape. Others may simply have gone the way of the dinosaur, leaving behind no descendants, not a single gene, just a few petrified scraps of bone—muddled messages from another epic that we haven't yet deciphered.[5] Our emergence was a far messier evolutionary business than many scientists suspected at first. Nevertheless, piece by piece, we are getting a handle on how we came to be the upright walking, naked apes we are.

The Human Family Tree

The fossil record is jumbled and controversial, and won't be resolved in this sidebar or even in this book. But it is helpful to have an overview of our family tree based on the fossil finds paleoanthropologists have made so far. The problem with the fossil record, of course, is that it is an incomplete puzzle. Though our picture of human evolution grows, in some ways, clearer with each new find, each find also raises questions. There have been many lines of primates that evolved at different times in different parts of Africa, each with their own distinct capabilities, intelligence, and anatomy. Some of these lines may or may not have interacted.

The general picture is this: In various parts of Africa, apes that had previously spent most of their time in the jungle began to stand upright. They probably did so haltingly and possibly for different reasons. Based on the most recent fossil evidence, the earliest primate to evolve into the line that ultimately led to us is *Sahelanthropus tchadensis,* a creature that lived in north-central Africa as many as seven million years ago. Some scientists disagree and argue that *tchadensis* may have been an early form of modern gorillas. The question has yet to be answered.

About six million years ago another primate, known as *Or-rorin tugenensis,* arose in what is now western Kenya. *Tugenensis* may have walked on all fours as well as two feet, and he may be related to us. The paleoanthropological community is still working that out.

Between 4 million and 5.8 million years ago two other controversial primates evolved in East Africa: *Ardipithecus ramidus* and *Australopithecus anamensis.* It is not clear how fully bipedal *ramidus* was, but there is little doubt that *anamensis* spent at least much of his time on two feet.

Beginning about 3.5 million years ago and continuing until 1 million years, there seems to have been a miniboom of new savanna apes, each species struggling to eke out a living on the increasingly naked plains of Africa. This might simply be an anomaly in the fossil record. There may have been more species earlier, but fewer of their bones may have survived. Or there may have been others who came afterward that simply haven't been found. Whatever the case, two species stand out: *Australopithecus afarensis* and *Australopithecus africanus.* They would have been a little taller than a chimpanzee, with both legs and arms proportionately longer than ours. If you were to come upon a group of either *africanus* or *afarensis* making their way through the southern African Transvaal or over a ridge in the African Rift Valley, you might mistake them for a troop of chimpanzees except for their upright posture. They were small, weighing in at 65 to 150 pounds and no taller than five feet. Unlike the gorillas evolving in Africa's central jungles, their limbs were thin and gracile. Their fossils tell us that they would have been smarter than the very first big-toed apes that emerged from the receding jungles like *Ardipithecus,* with brains that weighed roughly 450 cubic centimeters (cc), about the same size as today's bonobo chimps. There were also other gracile apes in the mix, though there is debate about which of them actually represent separate species and which are simply other forms of *A. africanus* or *A. afarensis.* These include *Kenyanthropus platyops* and *Australopithecus garhi.* (See chart on pages 6–7.)

While these more gracile apes roamed the African landscape, another group of bipedal apes were also evolving. They were taller, thicker, and stronger than *A. africanus* and *A. afarensis;* paleoanthropologists like to call them "robust." Their brows were sloped, their chests bigger, and their faces and jaws large, the better to house the square rows of chunky teeth inside, which were adapted for eating nuts, roots, and leaves harvested as they roamed Africa's woodlands and savannas. To support the enormous muscles needed to grind the food they ate, they evolved large sagittal crests, a thick row of bone that ran from the front to the back of their skulls and to which long jaw muscles were anchored. These creatures included *Australopithecus robustus, Australopithecus aethiopicus,* and *Australopithecus boisei* (also known as *Zinjanthropus boisei*). The brain sizes of these creatures vary widely, with most weighing about 400 cc.

Some scientists argue that the eating habits of these two groups of apes were a crucial turning point in their evolution. The more gracile apes may have been more prone to eat meat, mostly scavenged. Even partial meat eaters would have required a smaller digestive tract and less energy to digest their food. That may have resulted in two fundamental evolutionary shifts. First, the energy needed to operate larger digestive tracts may have been used instead to build bigger brains; and second, the more concentrated forms of protein in the meat gracile apes scavenged would have supplied the building blocks that accelerated their cerebral growth.

Even if *A. africanus* and *A. afarensis* were meat eaters, they were not feared predators; they didn't have the natural equipment for it. For the most part they were probably gentle, highly social, and above all, dependent on the other members of their troop for survival. And survival would have been a full-time job because life for these creatures could not have been easy. They had no real tools, no fire, no language, no claws or weapons. Their numbers would have been measured in the thousands, if that, certainly not the millions. Infant mortality would have been high and life spans short.

There is a good chance that at least some of these species interacted with one another. They may have interbred or they may have helped wipe one another out or they may have peacefully coexisted, like wildebeests and elephants, we don't know.

Whatever the case, the bigger, robust lines all died out, but the gracile apes evolved through one line or another into *Homo habilis,* the first of our genus and the first toolmaker (see chapter 3). Truthfully, the fossil specimens of *Homo habilis* could arguably be divided into more than one species, but for now they all tend to fall under the *habilis* nomenclature.

Most agree that *Homo erectus* is a descendant of *Homo habilis,* but their brains and body sizes are so different (some *erectus* specimens' brains are 50 percent larger and their bodies nearly a foot taller than *Homo habilis*) that it's likely other, still undiscovered species came in between. One candidate was found in 2002 in Dmanisi, Georgia, known fittingly as *Homo georgicus. Georgicus* was about the same height as *habilis,* five feet or so, but its brain was larger than most *habilis* specimens, weighing approximately 650 cc. This creature lived north of the Middle East about 1.8 million years ago, before *Homo erectus* reemerged from Africa and began fanning out all around the planet.

Following *Homo erectus* came a series of creatures that have shed light on our lineage but hardly paint a perfectly clear picture. *Homo ergaster* and *Homo antecessor* preceded *Homo sapiens neanderthalensis* (Neanderthal Man) and *Homo floresiensis,* a primate discovered on the Indonesian island of Flores in 2003. *Floresiensis* was clearly very advanced. It probably made tools and possibly spoke or used some sort of language, but stood no higher than three feet full grown, with a brain roughly the size of a chimpanzee's. (This seems to prove that it is the structure of the brain, not its size, that matters.) Current theory holds that *floresiensis* is a form of *Homo erectus* that evolved to dwarflike dimensions like other mammals on the island as an adaptation to the scarcity of food and the small size of the island's ecosystem.

Neanderthals are a different matter entirely. They were intelligent early *Homo sapiens,* but not, by common consensus, directly related to us. (Anthropologists debate whether our ancestors and Neanderthals might have mated.) In fact, our direct ancestor, Cro-Magnon Man, may well have wiped them out, probably because he had developed more advanced tools.

In between modern *Homo sapiens sapiens* and *Homo erectus* came another archaic form of *Homo sapiens,* known in most paleoanthropological circles as *Homo heidelbergensis,* a creature who had some of the facial features of *erectus*—a sloping brow, for example—but also some of our modern features, such as smaller teeth and a more rounded skull. They flourished between 500,000 and 200,000 years ago, when the first modern humans emerged.

Though fossils indicate that these earliest modern humans were physically identical to us, the wiring of their brains may not have been because it isn't until around 160,000 years later that we see the first flowering of complex artwork, sculpture, and other forms of truly modern culture and communication. Did disparate parts of a complex brain connect in ways that led to consciousness and insight that made us far more fully self-aware? That mystery still remains unresolved.

. . .

Under a searing noonday sun on November 30, 1974, Donald Johanson and his colleague Tom Gray were heading back to base camp for a break. Noon was a good time for a break because it was a poor time for fossil hunting in Hadar, Ethiopia. Not only was it hot, but the midday sun also threw no shadows, and the bone and rock all tended to blend into one indistinguishable morass of tan dust.

That morning neither Johanson nor Gray had turned up anything better than a few old pig and monkey bones. But then as they walked through a small gully just beyond the place they had been excavating, Johanson noticed what appeared to be an elbow joint lying at the bottom of a slope. Both men knelt down to look more closely when they suddenly realized that they were surrounded by hominid bones—the elbow, a femur, a chunk of pelvis, various vertebrae, and several ribs. As he picked through the rock, Johanson recalled fervently hoping that all of these fossilized messages belonged to a single human precursor. But he was afraid if he said so out loud, it might somehow jinx the possibility.[6] He needn't have worried.

Johanson called the creature they had found that day Lucy, for the popular Beatles song "Lucy in the Sky with Diamonds," but he named the species she represented *Australopithecus afarensis,* the southern ape from Afar. (The area where Lucy was found sits squarely in Africa's Afar Triangle, a chunk of land where three tectonic plates are still pulling that part of the world in separate directions.)

Lucy revamped the scientific view of human evolution and remains one of the most important paleoanthropological finds of the twentieth century because the bones Gray and Johanson uncovered that day told the world that Lucy walked upright at a time earlier than any scientist thought possible. Her pelvis, femur, and tibia all sent a clear message that despite being chimpanzeelike in size and stature (she was about three feet, six inches tall and weighed about sixty-two pounds), Lucy's view of the world was nothing at all like a chimpanzee's.[7]

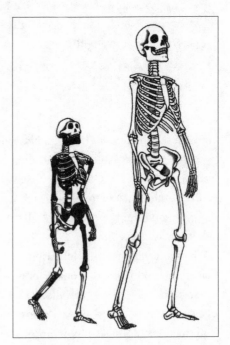

Lucy's skeleton (left) compared with the skeleton of a modern woman. Lucy's pelvic girdle was not like a modern chimpanzee's, nor was it fully human, but it was getting there. (Reprinted from *How Humans Evolved* (second edition) by Robert Boyd and Joan B. Silk, used by permission of W. W. Norton & Company.)

· · ·

At about the time Donald Johanson was assembling Lucy's bones, fabled anthropologist Mary Leakey (wife of anthropologist Louis Leakey) had returned to an area called Laetoli in Tanzania, a spit of dusty flatland about thirty miles south of one of her and her husband's favorite fossil haunts, Olduvai Gorge. Neither had been back to Laetoli for decades, and with good reason. Even in their earlier visits Laetoli had failed to yield any significant fossil finds. This particular expedition didn't turn up any bones of importance either. But it did produce the record of one of the most famous walks in history: Eighty feet of three sets of footprints perfectly preserved in a combination of mud and volcanic ash.

Initially, Leakey didn't think much about the discovery. The prints, found beneath a relatively thin layer of dirt, were interesting at first glance but not apparently very important. They looked as if they had been made a few thousand years earlier by three people strolling through the gorge. Leakey figured that Mount Sadiman, once an active volcano that loomed nearby, had probably spewed the ash that had created a kind of cement that preserved the walkers' journey.[8]

But then in 1976 Leakey finally got around to having the rock containing the prints dated, and she was astonished to find that the walk had been made not a few thousand years earlier, but 3.5 million years ago, long before any modern humans roamed Africa. Or put another way, the footprints revealed that a species, probably like Johanson's Lucy, walked upright with a fully human gait. *That,* Leakey admitted years later, was when "we got excited!"[9]

"Make no mistake about it," Timothy White, an anthropologist who worked with both Johanson and Leakey later said, "[the Laetoli footprints] are like modern human footprints. If one was left in the sand of a California beach today, and a four-year-old was asked what it was, he would instantly say that someone [human] had walked there. He wouldn't be able to tell it from a hundred other prints on the beach, nor would you."[10]

Leakey's discovery pushed the date of bipedal hominids back even further than Lucy had—four hundred thousand years further. Before these discoveries no one had dreamed that upright-walking apes could have lived this long ago. But all of the evidence was there in Lucy's bones and in the footprints Leakey had uncovered: the slender foot with its knobby big toe

Laetoli footprints preserved in Tanzania (left); close-up of a footprint (right). These footprints are at least 3.6 million years old, yet they are nearly identical to our own. (Used by permission of Science Source Photo Research, Inc., in New York, New York.)

supporting the weight of the smallish, long-armed bodies, pushing them away from the volcano and toward their destination. The heel was already elongated, the toes ran in parallel, and the arch was engineered and in place, absorbing the weight and transferring it along the outside, then across the balls of their feet, just as our feet do today.

Imagine the haunting image of these three creatures—two adults and a child, by the looks of the prints, all unlike any primate that lives on Earth today—passing together through a flat spit of land. Behind them loomed Mount Sadiman, rumbling, belching hot ash, rattling the ground beneath their feet. Yet they weren't running scared. Perhaps they were accustomed to irascible volcanoes. The footprints are measured; there is no sign of flight. In fact, the impressions of their feet show that one of the creatures stopped briefly, turned to look eastward at the angry volcano, and then continued onward.[11]

No one can say precisely what these three creatures looked like. Apparently they survived their journey and went on to live out their lives among the shattered and restless landscape of the Rift Valley, so we don't have their remains. But if they had the same anatomy as Lucy, and most scientists believe they did, from a distance their walking, if not their bodies, would very likely have resembled the gaits of a toddler and two adolescents making their way through a park.[12] Their hips would have twisted like ours, and their arms, though slightly longer than a modern human's, would have swung in a very human way. Despite their small stature, their legs would have angled inward, unbowed, and anchored to a slender pelvis. From the hips down they would have looked remarkably similar to us.

. . .

Johanson's and Leakey's discoveries turned theories about human evolution on their heads. Until Lucy, scientists were mostly certain that if anything distinguished our ancestors from their simian cousins it was their brain, not their feet. Brains, they theorized, led to bipedalism rather than the other way around. But apparently they had it wrong. Lucy was not big-brained, at least by our standards. The skull fragments that Johanson and his team found indicated that her cerebral capacity weighed in at about 450 cc, roughly the same as a modern chimp. Yet Lucy's locking knee joint and short, narrow pelvis indisputably indicated that she stood upright.

The footprints at Laetoli sent the same clear message. Our ancestors had begun walking upright sooner than we thought, perhaps a million years after our line split off from the common ancestor we shared with chimpanzees. In an evolutionary blink, they had gone from knuckle-walking, tree-climbing jungle apes to striding, sure-footed, savanna apes that walked pretty much the way the rest of us do today.

This was remarkable and puzzling. What, after all, would have caused these creatures to stand upright so quickly? And how did it happen?

. . .

In *Gray's Anatomy* the curious appendage we carry at the end of our foot is called the hallux magnus. Most of us know it as our big toe. It is an odd-looking thing that we generally take for granted. But we shouldn't, because had our predecessors never developed their big toes, they would never have stood upright. And had they never stood upright, we would not be here to ask how such a thing could have happened in the first place.

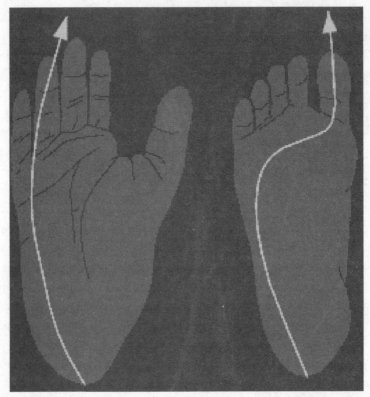

A comparison of the shape of a chimp foot (left) and a human foot (right). Ape feet look remarkably similar to human hands. In a human's foot, weight travels along the outside and then shifts across the ball of the foot to the big toe. (Used by permission of WGBH in Boston, Massachusetts.)

Like modern chimps, gorillas, and orangutans, Lucy's predecessors had no experience of big toes, at least not our variety. Their inside toes would have been more like crude thumbs made for gripping branches rather than for pushing off hard ground.

Gorillas and chimpanzees can walk upright, but they don't do it very well. Their pelvises and legs push the weight to the outside of their flat feet, and they tend to roll from side to side. Five million years ago, when the first savanna apes and their jungle-dwelling cousins began to head off in separate evolutionary directions, they very likely shared the same bowed legs and big, square pelvises. Their feet would have looked nothing like human feet, but remarkably similar to human hands. And the four outer toes would have been considerably longer than ours.

But the real difference between these predecessors and modern humans would have been that inner toe, which remained set apart from the others like an unwanted dinner guest. At its base it would have turned outward and then run back in at the top, adapted for grabbing and holding.*

The traditional view is that the ancestral apes from which we sprung began to benefit from the evolution of a big toe about five million years ago. It was then that their crescent-shaped, thumblike digit would have begun to drift inward, where it grew knobbier and less fingerlike, a trait that eventually enabled it to support 40 percent of its owner's body's weight.

At least that's the way Darwin would have imagined it. For him all evolutionary change came gradually. But the fossil record doesn't always support Darwin's views. Stephen Jay Gould famously brought this to the world's attention with his theory of "punctuated equilibrium"—the idea that big genetic changes can occur in relatively short periods of time. Gould pointed out that occasionally species seemed to make sudden, apparently inexplicable leaps in their looks or anatomy, as if some evolutionary switch had been flicked.[13] When it comes to the evolution of the big toe, that may be precisely what happened.

The most dramatic example of punctuated equilibria in all of evolution is the Burgess Shale, an ancient ridge of rock about a city block long discovered one day in 1909 by a young paleontologist name Charles Doolittle Walcott when he was horseback riding in British Columbia. Walcott realized he had found something amazing, and between 1910 and 1925 he mined 80,000 specimens from the ridge. Among the rock he found no fewer than 140 new species of ancient creatures. But more importantly, his discovery revealed that about 500 million years ago, at the dawn of the Cambrian period, life suddenly blossomed into a shocking number of forms—trilobites, brachiopods, ancestral starfish, sea urchins, and alien creatures like *Opabinia,* with its five eyes and fire hose nose. Whatever their forms, the creatures seemed to have emerged from nowhere. In one layer of the fossil record they didn't exist. In the next they were everywhere.

Gould put it this way: "The Burgess Shale included a range of disparity in anatomical designs never again equaled, and not matched today by all of

---

* If you doubt the importance of the hallux magnus, try walking with it raised off the ground. If other toes are injured, we can get by, but robbed of our big toe, we not only have severe problems walking properly, but, as any NFL running back suffering from "turf toe" can tell you, it becomes impossible to run or jump or cut rapidly and gracefully in one direction or the other.

the creatures in the world's oceans." The foundations for the body design of every living animal (and then some) that ever lived afterward, Gould concluded, had evolved in this evolutionary blip. The evidence sat right there in the ancient, hardened muck of the Burgess Shale.

Though scientists now know that all of these creatures did not instantly appear as if by magic, they did arrive rather suddenly, considering that for the better part of the previous three billion years all that life on Earth had been able to manage were single-celled bacteria, plankton, and a few species of multicelled algae. Gould, and others who have studied these sudden evolutionary leaps, could not explain how they happened, only that they apparently did. But then in 1984 American scientists Mike Levine and Bill McGinnis were studying fruit-fly embryos when they discovered what are known as HOX genes, and a possible explanation began to unfold.

HOX genes are "master switches," which toggle other strings of genes on and off. Since McGinnis's and Levine's discovery, other scientists have found them in every existing animal, including humans. In the fly embryo, HOX genes control the number and length of body sections, including where, precisely, features such as wings, legs, and antennae end up. In humans and horses the same genes ensure that appendages like toes, heads, and feet take shape in the right location. But more importantly HOX genes, in whatever form of life they are found, control whether other, more specific genes switch on at all to form various appendages—arms or antennae or wings.

Since HOX genes control squads of other genes, the results can be dramatic when they mutate. Recently scientists have found that several human malformation syndromes can be caused by an alteration in a HOX. When a developing embryo's *Hoxd 13* gene mutates in a certain way, for example, a condition can result in a baby being born with more than the normal number of fingers or toes. Doctors call this polydactyly syndrome.

But while in the medical world afflictions such as polydactyly may be considered "malformations" and "conditions," an evolutionary biologist might just as easily call them genetic mutations. But the thing is, they don't develop gradually. They appear instantly.[14]

Some scientists believe that HOX genes can help explain Gould's punctuated equilibria. And in particular some believe that their existence may help explain how we managed to go from knuckle-walking jungle apes to upright-walking savanna apes in a few hundred thousand years or less.[15]

Scientists generally agree that most evolutionary adaptations are gradual, just as Darwin theorized, but some believe that they can show up inside of a single generation. If those adaptations improve a creature's chances of survival, they would tend to spread through the rest of the gene pool. And their spreading would more likely happen when an environment has radically changed because that would make a dramatic mutation more likely to serve a purpose.

In the case of our ancestors, perhaps their DNA shifted and a gene like *Hoxd 13* mutated and reshaped the feet of a particular family of savanna apes more than four million years ago that enabled them to more easily walk upright. In another time and place an australopithecine doctor might have written the alteration off as an unfortunate syndrome, but given that our ancestors found themselves struggling to survive in open grasslands that were nothing like the dense jungles in which they had originally evolved, a differently shaped toe might have been precisely what the doctor ordered. And it may explain how we got up and running so soon after we found ourselves orphaned among East Africa's expanding savannas.[16]

. . .

Once our ancestors' knobbier big toes emerged, nature clearly favored other mutations that enabled walking and running. The remaining eight toes contracted, for example. Heels grew longer and thinner, and an intricate system of small bones and muscled arches developed that absorbed the constant shock of body weight being transferred incessantly from left to right and back.

In time fully one quarter of the bones in the human body came to be located in our feet, and our big toe developed enough strength to support 40 percent of our weight.* When we run or jump, this system, woven together by a web of 141 tendons, muscles, and ligaments can handle 6,000 pounds of pressure without a sprain, all of which helps explain the twists, turns, leaps and vaults that amaze us when we have watched athletes and dancers seemingly fly through the air and land like feathers. Other primates may swing through the trees with the greatest of ease, but none can run, stride, or leap as gracefully as we do.

Nevertheless, our smooth gaits are not solely the result of our finely

---

* We have twenty-eight bones in each foot if you count the sesamoid bones at the base of the big toe.

engineered feet. Other anatomical rearrangements also followed, bringing still shinier clusters of metamorphoses in tow. These did not likely happen neatly in a row. They probably arrived in fits and starts, with one adaptation feeding back on another. Some species of early savanna apes may have been more upright than others. Some may have developed knobbier toes, while cousins evolving two mountain ranges to the east or west did not. We can't know all the details. There was evolution and coevolution and reevolution, mystifying exchanges between changing environments and the accidentally transformed strands of DNA that arose from one troop to the next.

These environmental conversations led to four other broad changes in the hominid line that eventually enabled savanna apes to walk and run very much the way we do today. First, our bowed legs straightened. Lucy's remarkably preserved skeleton shows us that. Her legs, though not as long as ours, had already begun to bend inward from the thighs to make her the most knock-kneed primate the world had seen yet. She also developed stronger gluteal abductors, muscles on the sides of her hips that contracted to prevent her body from toppling left or right when all her weight shifted from one foot to the other in midstride.

Second, the pelvis and hip joints morphed. A chimp's pelvic saddle, the circle of bones that connect the legs to the torso, is longer and straighter than ours, and the legs are hinged to its side at something like a ninety-degree angle. While this is fine if you are bent over at the waist most of the time walking on all fours, it is a problem when standing upright. Simian hip bones have the shape of shoehorns, and their angle is perfectly vertical. But to balance their upper bodies less precariously on top of their legs, the pelvises of hominids like Lucy had to grow shorter and spread out so the hip joints came to connect at something closer to a forty-five-degree angle. Lucy's pelvis, strangely enough, is even more splayed than ours. Her hips are shorter and more open. This may be because she was lighter and didn't require the same amount of bone to support her upper body. But whatever the reason, Lucy's pelvic girdle was nothing like a modern chimpanzee's. Nor was it fully human, but it was getting there.

Third, as weight was hoisted up on top of their pelvis, our predecessors' spines realigned. The spines of chimps and gorillas are straight. They can afford to be because most of the time they are bent over with their backs parallel to the ground. This means that they are carrying about half their weight on their feet and half on their knuckles. Our spines (and the spines of *Australopithecus afarensis* and *africanus*), on the other hand, bend in the shape of an S, inward at the bottom and then outward as the vertebrae run up to the neck, a piece of evolutionary engineering that more effectively handles the weight that arms and hands had once supported.

Inevitably these reconfigurations shifted the location of our heads, something that eventually had immense consequences. When you are at the zoo and a gorilla walks toward you on all fours, his head is tilted back so he can look in front of him. If you had X-ray vision and could see the vertebrae of the gorilla's neck, you would notice that they enter at the back end of the base of his skull. This makes anatomical sense if you normally walk on all fours, but doesn't work very well if you stand upright because then you would find yourself staring at the sky. For an upright-walking ape, this arrangement had to change.

The place where the head and neck connect is called the foramen magnum. If you sat a gorilla and a human side-by-side and looked straight down through the tops of their heads, you would see that while the spine of a gorilla enters at the back of the cranium, ours enters at the middle, so that our heads tilt forward and sit squarely on top of our bodies, making it the final piece of skeletal statuary to be stacked in one straight line from the tips of our toes to the tops of our necks.

This was a lot for a readapted big toe to accomplish, but fortunately for us it did because there were enormous advantages to standing upright in Africa's new open and dangerous woodlands.[17] When the sun fell away on a moonless night, the darkness that followed would have been more black and limitless than anything we can imagine today in our artificially brighter world. At any moment our ancestors could have been snatched away by a nighttime predator. Often they probably were. No wonder we still worry about things that go bump in the night.

Life in broad daylight wouldn't have been much easier. In the dry season, with the equatorial sun beating down, temperatures would often have reached triple digits (Fahrenheit). Troops of apes would have been kept constantly busy gathering food, finding water, and taking care of the young. On the savanna they would have had to contend with new predators that were evolving just as they were, early versions of jackals and hyenas and the megantereon, a lion-sized, saber-toothed cat with teeth the size of daggers. Though australopithecines could run faster than their knuckle-walking cousins, they could never hope to match the speed of predators like these.

Considering the circumstances, the remaining trees that clustered here and there along riverbanks and lowland mountain ridges must have been welcome sights, a kind of hominid security blanket they could recall from their simian days. Apparently they made good use of them. The fossil record reveals that Lucy and her cousins were still first-rate climbers, even if they walked upright. They had long, apelike arms, and wrist sockets that snapped into place for swinging from tree branch to branch when the situation called for it.

But standing on two feet was the adaptation that helped most. It vastly improved any savanna ape's chances of survival. Knuckle-walking may have been ideal for short forays through the dense jungle undergrowth, but in East Africa's expanding grasslands it would have been slow, tiring, and ultimately deadly.[18] Studies show that knuckle-walking chimpanzees use up to 35 percent more energy than we do when we walk upright. That would never do in the open woodlands of the late Pliocene and early Miocene epochs.

Today's surviving tribes of hunter-gatherers, like the bushmen of the Kalahari, are living proof. Studies have shown that each day they typically cover six to eight miles tracking down the food they need to survive. Had savanna apes been forced to do this walking on their knuckles, not only would they have used a third more energy (and a lot more time) to locate their food, they also would have had to eat a third more calories to replenish the additional calories their knuckle-walking required. Standing upright was the best way to survive. Maybe the most convincing proof is that today there is not a single knuckle-walking ape still roaming Africa's grasslands.

Though food in the savannas was undoubtedly tougher to come by than it was in the jungle, there would have been some sources. Fresh prey, as well as nasty predators, were evolving on the grasslands: the deinotherium, for example, a thirteen-foot-high elephantlike creature with down-curved tusks; a giant buffalo called a pelorovis with a horn span six feet from end to end; and an early form of today's giraffe that was short-necked and a mere seven feet tall. These would have supplied our upright-walking ancestors significant supplies of protein as they scavenged carcasses left behind by the grassland's big cats.[19]

It is even possible that early hominids did some hunting and killing of their own. Standing upright, after all, not only meant getting around faster, it also created a taller creature that could now throw objects with more force and accuracy. Chimpanzees have been shown to hunt small game; perhaps *A. afarensis* and *A. africanus* did as well.

We can't say for certain how many of these evolutionary scenarios are truly accurate, but paleoanthropologists have invoked them all to help explain how and why the first savanna apes did the supremely odd thing of developing toes that enabled them to rise on their hind legs. All of them, to some degree or another, are probably true, and the last possibility might be the most compelling because it means our predecessors could not only avoid becoming a meal, they were also able to improve their chances of finding one. One way or another, standing upright would have helped solve the number one challenge every creature must solve if it hopes to avoid death and successfully pass along its genes: live to eat another day. For our ancestors, bipedalism became their version of sharp claws or deadly fangs, a radical adaptation that enabled them to survive in their equally new environment.

But upright walking accomplished even more than that. It changed the way we looked, too, and it changed how we saw the world and behaved, mostly with one another, in deeply fundamental ways. In other words, it not only shaped a new kind of body, it also shaped a new kind of mind.

Chapter 2

Standing Up: Sex and the Single Hominid

The evolution of sex is the hardest problem in evolutionary biology.
—*John M. Smith*

A NIMALS AVOID EXTINCTION two ways. First, they adapt to their environment. Second, they compete among their own kind for the affections of the opposite sex. Darwin called this intraspecies competition "sexual selection," and he wrote extensively about it in *The Descent of Man*. His point was basically this: To survive in their environment, all animals have to first outflank predators, disease, dangerous weather, and whatever else nature throws at them. Those that develop the right physical equipment to live another day pass their well-adapted genes on to the next generation.

But to pass their genes along, they also have to successfully procreate, and to manage that, they have to gain the attention of the opposite sex. In the case of males they may evolve brightly colored feathers, a lion's mane, or an elk's fourteen-point crest of antlers. Each is a way of saying, "Look here! I have great genes." These features, Darwin pointed out, "acquired their present structure not from being better fitted to survive in the struggle for existence, but from having gained an advantage over other males."[1]

The point is that in the never-ending struggle to win over the animal they want to mate with, species develop some interesting behaviors and physical traits. Our early ancestors were no exception.

Our sex organs are unusual, for example. Take the human penis, which, to be blunt, is enormous compared with its peers in the rest of the primate world. Unlike humans, if male gorillas, orangutans, chimpanzees, and bonobos aren't sexually excited, their members are so small and hidden from view you would have no way of knowing they have any at all.

The same applies to human female breasts, which are large and defined compared with other primates. There is no obviously practical reason for this. Their increased size isn't, as you might expect, necessary for lactation and feeding. All other female primates suckle their young but never develop full, round breasts.

Our buttocks are a third, strangely human adaptation. We are the only primates that have a round, muscled bum. Arguably it developed when we stood up because we needed those muscles to help support and balance all the weight we were now carrying above our pelvises. But they may also have evolved as a sign of health that made us more attractive to the opposite sex. In men, solid, rounded buttocks might have signaled the strength needed for running and hunting, traits that make a good provider. But in females the dynamics may have been different. Scientists have found that males across all cultures rate women as consistently more attractive if the circumference of their waist is 70 percent of their hip circumference. The theory is that the unconscious, primal message an hourglass figure sends is that the owner of such a body is healthy and fertile and therefore desirable.[2] This may explain why women as different as Marilyn Monroe, Twiggy, Sophia Loren, Kate Moss, and the Venus de Milo are all considered attractive. Every one of them has a waist-to-hip ratio of about 0.7. Apparently if size doesn't matter, at least sometimes ratio does.

However we calculate the mathematics of human sexuality, the point is that these peculiarly human features evolved not simply because they helped us to survive but because they helped us to multiply. They emerged because they enabled our ancestors to better compete among themselves to find an appropriate mate so they could pass along their particular set of genes. Or put another way, they made us sexy and attractive.

In his book *The Prehistory of Sex*, Timothy Taylor theorizes that human males evolved large penises to show the males that they were competing with and, possibly, other females, what they had to offer. It became a signal of manliness and fertility. The issue wasn't so much what the penis could do—ejaculate and impregnate; all penises do that—it was what it symbolized.

A visible penis was a reminder of its purpose, a sexual cue. But if that is the only reason for evolving large penises, why did our ancestors begin to develop them when apes didn't?

Taylor holds that upright walking made the penis more visible and that, in turn, made its symbolic meaning easier to communicate. There it was for everyone to see, even when it was "at ease." Not only that, but if locker room behavior is any indication, it also became a source of bonding among males because it was an obvious and unique sexual feature common to all of them. Despite the old saw that "size doesn't matter," Taylor argues that at least some females from generation to generation must have preferred bigger over smaller; otherwise the "bigger" genes would never have been passed along that eventually led to increased size being a common human trait.[3]

The same may be true of female breasts. Perhaps they came to serve a symbolic meaning similar to penises, although understanding how that happened requires following a particularly circuitous, evolutionary path.

Many primate females, but not humans, have what is known as estrus skin between their legs that engorges and becomes much more visible when they are fertile. In evolutionary terms it's a straightforward way of saying, "I'm ready when you are to continue the species." During these times sex can, like voting in Chicago, happen early and often.

But once savanna apes stood up, estrus skin wouldn't have been nearly as noticeable, hidden as it became between the two legs of an upright body. If our ancestors were simultaneously developing larger, more muscular buttocks, then the estrus skin would have been even less obvious, and that would have opened the door to other adaptations.

There is a theory, for example, that sometime in our past, as we were in the process of standing upright, we began, slowly, to grow less hairy. Darwin believed this happened early in human evolution, but there is no absolute way we can be sure. For the sake of argument, though, let's assume it did. (As we will see, there is reason to believe it happened at least two million years ago.) If estrus skin was hidden, and if buttocks were just beginning to evolve, scientists speculate that two changes may have occurred. First hominid bottoms may have grown more muscular to power our new form of locomotion, *and* more fleshy to store body fat, something crucial to survival in an environment where it was never clear where the next meal would be coming from. In females, this fleshiness may also have substituted for the engorged estrus skin that had become less visible.

Zoologist Desmond Morris speculated in his book *The Naked Ape* that our ancestors' bottoms would have become naked as a way to help this new version of sexual skin to stand out from other, more hairy parts of the body. A hairless, rounded rump would have indicated it belonged to a female with a certain level of fat, and a certain level of fat would have been attractive because it was an indicator of health and fertility. We know today that women need to have minimum amounts of fat in their systems to ovulate. Studies reveal that some long-distance female runners who have very low fat levels sometimes stop menstruating and ovulating. Other studies show the same happening to women who are on extremely low-fat diets.[4] So a rounder rump would have been a solid indicator that this female not only had a worthy set of genes, but also was fertile enough to pass them along.

All of this circles back to the evolution of female breasts, but again the route is indirect. Species sometimes develop sexual skin on one part of the body that reflects or recapitulates sexual skin on another part. Female gelada baboons, for example, show off brightly colored patches of skin around their nipples that look very similar to the estrus skin that flashes on their bottoms. If they are sitting down, which is something they do a lot, this gives them a second way to attract sexual attention. Mandrills also have ribbed, brilliant blue, white, and scarlet muzzles that recall the colors of their genital area, a kind of natural neon sign advertising their sexuality.[5]

Full, round female breasts might have evolved as a recapitulation of their newly developed bottoms. Once we stood upright, chests, which are mostly hidden on animals that walk on all fours, were now front and center. Because our ancestors had gotten up on their hind legs, they had more upright, face-to-face contact. For females, their chests would have made perfect billboards for recalling the round, fleshy buttocks that had recently evolved to symbolize womanhood, health, and fertility without compromising the original purpose of their breasts.*

We can't be sure that these particular features evolved this way. Fossilized bone isn't very helpful here, and other factors were certainly part of the evolutionary mix. The heat of the savanna undoubtedly played a role in our nakedness, and the need to capture and share certain scents and odors for

---

* The same forces may have been behind the development of the full, red lips we humans have, which may recall female genital labia. Some scientists theorize that beards on men mark a recapitulation of their genital area as well.

sexual purposes may help explain why we haven't become entirely hairless. Eventually some hominids, between three million and two million years ago, developed, and wrote into their genetic code, sexual signals and features that found their way into the long chain of DNA that shapes us today. We have bottoms, breasts, and penises, and their effect on our psyche, culture, and behavior remains large. They send powerful sexual signals, even in modern society, and they are crucial erogenous zones and pleasure centers that continue to shape the way we act and what we value.

At this point in their evolution, the hominids from which we descended must have felt like adolescents caught in the grip of raging hormones, sexual desire, and social turmoil. Here they were, struggling to survive from one day to the next on the grasslands of Africa, battling predators, disease, and the elements on the one hand, and one another on the other. Although standing upright had undoubtedly saved the species, it also had introduced new physical, social, and sexual forces that made life more complicated. And still there was more. All of the anatomical realignments our big toes

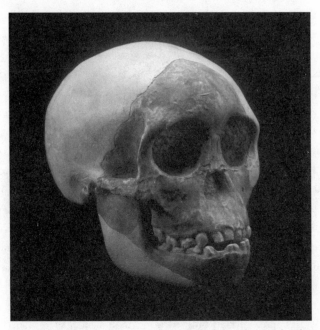

The Taung Child skull. Because of his youth, Taung looked more human than his apelike parents, the result of a strange phenomenon in nature called neoteny. We may be neotenous chimps. (Used by permission of the Smithsonian Institution.)

had set in motion were creating yet another evolutionary bottleneck that threatened to strangle the species.

. . .

In 1924, workers from a South African limestone quarry handed Raymond Dart, a young anatomy professor at the University of Witwatersrand in Johannesburg, a peculiar skull. The laborers had sifted it out of the debris of dynamited lime deposits in their home, a village called Taung, on the far-away edges of the Kalahari Desert. Lime deposits in this part of Africa were already well known for their fossils—mostly baboon skulls and ancient animal limbs—but when Dart saw this particular two-and-a-half-million-year-old skull, he knew he was looking at something no one had ever seen before. And it was not a baboon.

What amazed Dart was how human the skull looked. It was small, but its cranium was rounded and unusually large compared with the rest of the face. There was no thick brow ridge, as one might expect in a primate that was so ancient, and the forehead was vertical from the nose up. He also noticed that the foramen magnum, that spot where the spine enters the skull, was far closer to the center than he had seen in any other primate except humans. Such a creature would have walked far more upright than chimps and gorillas.

Dart named his find *Australopithecus africanus* (the southern ape from Africa) and concluded that he had found "an extinct race of apes intermediate between living anthropoids and man." Still, that didn't explain its perplexingly human form. Almost no other fossils of human predecessors had been found up to this time, so scientists could only conclude that this was simply the way protohumans must have looked. But in this, they were wrong.

What Dart had done, without knowing it, was step back in time twenty-five-hundred millennia and then glimpse the future. His discovery would not only provide new insight into our ancestry, it also would provide important clues about why we look, and are, so different from other apes.

The reason the primate skull that Dart had been handed included so many thoroughly humanlike features was because it was a baby, a toddler, perhaps two or three years old. Dart called the fossil the Taung Child. Two puncture marks in his head indicated Taung had met his fate when a leopard, or possibly an eagle, carried him off and made a meal of him. But had this child survived and grown up, he would not have looked nearly as human as he did the day he died. As he matured his jaw would have grown larger and more snoutlike, his eyes would have risen up in his face to be hooded by a bony brow ridge, and his forehead would have sloped back apelike from his nose.

But why should a hominid toddler that lived more than two million years ago look more human than a fully grown adult? Because adult humans resemble infant, even fetal, apes. The reason we do is because, in some ways, that is precisely what we are. The Taung Child's resemblance to a modern human is the result of a curious evolutionary phenomenon called neoteny, which has over time nudged us into being born earlier in our development than other primates. The result is that we tend to retain many of the youthful physical and behavioral traits of our species (and our ancestors) well into adulthood. In short, we are born younger, and stay that way longer. And the reason we are born younger is because it has evolutionary advantages, and because of our big toes.

. . .

A human being's entry into the world is the most dangerous and difficult in nature. Because standing up rearranged the pelvic bones of our ancestors, and because of the rather large dimensions of our heads, human babies have to rotate in the birth canal from facing forward to facing sideways as they begin to emerge. Then they have to turn still another ninety degrees until they face their mother's back as they are born. If the baby rotates the opposite way, the sharp bends in the birth canal could twist the baby's spine backward and do enormous damage.

For gorillas and chimpanzees the process isn't nearly as difficult. Monkeys deliver their babies through a considerably broader birth canal while squatting or on all fours. Because their heads are smaller, infant chimps, gorillas, and orangutans can exit facing the stomach of their mothers, and even help pull themselves out of the birth canal. As these babies do a controlled fall, mothers often reach down and guide the infant out.

The twisting, turning route we humans take was already in the works as far back as four million years ago. Fossils as old as Lucy's reveal that standing upright had narrowed the birth canal enough that even for relatively small-brained australopithecines, babies would have had to rotate either forward or backward to fit their shoulders through the narrowed canal. Karen Rosenberg at the University of Delaware and Wenda Trevathan at New Mexico State University, two experts in the evolution of birth, speculate that this also would have drawn the troop more closely together because pregnant australopithecines would have needed help bringing their young into the world.[6]

If birth was growing more difficult for Lucy and her kind, it would only have grown considerably more difficult for the next major species of savanna primate to emerge—*Homo habilis,* the first of our genus and our first known direct ancestor. *Homo habilis* (the handy man) walked out the mists of time about two million years ago with a brain that had nearly doubled, to an average size of 750 cc.[7] That made an already tight and dangerous journey into the world considerably more difficult. The birth canals of *Homo habilis* could not grow larger to accommodate the larger heads of their children. If they did, the upright walking that the big toe had made possible would have become impossible. Hips would have grown too wide and bipedalism would have become mechanically unworkable. On the other hand, going back to smaller brains was not in the cards. Our ancestors faced a dilemma. The same evolutionary forces that were making them more mobile and more intelligent were also making birth more difficult. Without a solution, increasingly intelligent, upright walking apes would become extinct. Something would have to give. And, fortunately, something did.

While developing in the womb, the skulls of primates don't form in one piece. They develop instead as separated plates that drift over the brain. Humans have eight plates, each of which over time knits together after birth to form the hard, brain-protecting shield of our cranium. Before birth these disconnected plates make our skulls pliable enough to slip through the tight circle of the birth canal.

Chimps and gorillas have some unknitted skull plates, so it is likely that our ancestors did as well. They were an elegant solution to a nasty evolutionary problem, and a perfect example of neoteny at work. It meant that *Homo habilis* didn't have to develop a larger pelvis or smaller brains, or die off. Instead their young simply entered the world premature. And the larger the brain, the more premature their births had to be.

We humans are the most extreme case of this in the primate world. If we were born as fully formed and physically mature as the babies of contemporary great apes, human gestation would last not nine months, but twenty-one! This means we are born a full year premature. *We* may define

A young chimp (left) and a mature chimp (right). Humans have retained many of the physical traits of young chimpanzees. (Reprinted from *Ontogeny and Phylogeny* by Stephen Jay Gould.)

"full term" as nine months in the womb, but by ape standards we are fetuses that have arrived twelve months earlier than we should have.

Because of our premature birth, we come into the world almost totally helpless. Our brains are small and underdeveloped; our limbs, fingers, and toes are cartilage rather than mature bone. We are born nearly blind, our nervous systems are not even close to fully formed, and we continue to grow for approximately a third of our lives, years after other primates have reached their majority. And while most of the plates in our skulls knit within the first several years of life, some don't close until age thirty, and in a few cases can remain open when we are past ninety years of age.[8]

All of the evidence suggests that unlike other species, our brains never stop adjusting to the world around us. Recent studies confirm the unusual plasticity of the human brain, and new evidence reveals that, contrary to popular belief, it does in fact replenish certain types of its own cells. Other research has shown that the prefrontal cortex, the most recently evolved part of our brain, continually rewires itself in response to new experiences until the day we die.

The way Stephen Gould saw it, this lifelong youthfulness represented a powerful form of evolutionary selection that enabled humans to enjoy the benefits of a highly adaptive brain long after birth. Jacob Bronowski called these neotenous qualities "the long childhood." Rather than binding our behavior irretrievably to our genes, they make us adaptable and mentally nimble. They enable us to change our personal behavior in reaction to our personal experience, and as a species, they keep us curious, playful, creative, and restless; in a word, youthful. And that youthful exuberance is, to a large degree, the bedrock of human culture and the creativity it represents.

. . .

Louis Bolk, a professor of anatomy from Amsterdam, first saw how these changes seemed to apply to humans, at least physically. Bolk was a great believer in neoteny, and in 1926 he assembled a list of features he saw in fetal and baby apes that also seemed to show up in the anatomy of human beings: a less snoutlike lower face, for example, our higher foreheads, and the greater ratio of brains to body mass. He pointed out that we are largely hairless like baby apes and that ape fetuses had external ears more like ours, a thinner facial bone structure, a foramen magnum more centrally located, and, strangely enough, a straight big toe.[9,*] "Man," Bolk observed, "is a primate fetus that has become sexually mature."

Bolk's concepts weren't universally accepted at the time, but as Gould later pointed out, he was clearly on to something. Because we are born with skullcaps that are soft and in pieces, we can not only bring larger brains into the world, but also grow them larger once they have arrived. When a chimp is born, its brain is already 40 percent of the size it will be when it is fully grown. And while it continues to develop after birth, it soon reaches its full size. We, on the other hand, are born with a brain that is less than a quarter of its full size (23 percent, to be exact). During the first three years of life it trebles in size. But that still leaves nearly a third more brain growth ahead of us, something that continues into early adulthood.

An interesting aspect of neoteny is that it doesn't change existing gene

---

* For example, as chimpanzees form in the womb, their big toe is initially straight, not unlike the one at the end of your foot. Later it curves so it can better handle tree climbing. But as our ancestors began to move into the savanna, any newborns that accidentally came into the world with the birth defect of straight big toes would have actually enjoyed an evolutionary advantage, and that advantage might have been passed along.

expressions, it simply postpones them. In our case by delaying or discarding the expression of certain genes, it drew a new sort of creature out of the womb into the real world. And when it did, it radically changed the way our kind subsequently evolved.

Are We Infant Apes?

During the first half of the twentieth century, several scientists added to Bolk's original list of human neotenous traits. Evolutionary biologist Stephen Jay Gould compiled an expanded list in his book *Ontogeny and Phylogeny,* published in 1977. The general consensus among scientists now is that Bolk's observance of neoteny was accurate, but his explanations for why and how it came about were not. (He believed it was due to a kind of endrocrine-induced retardation.) But it is obvious that we retain many features seen in infant and toddler apes in human adulthood, and the fossil record reveals in *Australopithecus africanus, Homo erectus, Homo habilis,* and *Homo sapiens* a progressive retention of juvenile traits. For example:

- Flat-faced orthognathy, the phenomenon that makes human brows less sloped and human jaws less pronounced than the jaws and brows of mature apes, but quite similar to the faces of infant apes.

- The lack of body hair. Newborn and young apes have less of it.

- Ear formations. Baby apes have ears that look more like ours.

- The central position of foramen magnum. This shifts backward as apes grow older.

- High relative brain weight. Toddler apes have a brain-to-body-weight ratio closer to ours than grown apes do.

- Persistence of cranial sutures. Other primates are born with sutures, but they close up much more quickly than ours do.

- Structure of hand and foot. Fetal apes actually have a big toe and foot shaped much more like human ones. Apes' curled big toes develop as they grow to adulthood.

- Absence of bony brow ridges. Young gorillas and chimps don't have bony brow ridges. They develop them later in life.

- Absence of cranial crests.

- Thinness of skull bones. We have hard heads, but they aren't as hard as the heads of other primates.

- Wider head.

- Smaller teeth.

- Later eruption of teeth. This is another way of saying that we remain toothless longer and in this way resemble apes, who are born without teeth.

- Prolonged period of infantile dependency. We are babies longer than other primates.

- Longer life span. Another way of saying we remain young longer.

- Prolongation of fetal growth rate. Humans remain in the womb longer.

Neoteny was the evolutionary equivalent of the wheel or fire. Once our predecessors could walk upright, grow increasingly intelligent, *and* still manage to survive birth, whole new worlds opened up, and whole new evolutionary forces fell into place. We were mobile and could range across the savanna, hunt and forage by foot far more quickly than our knuckle-walking cousins. We could apply our increasing intelligence to the problems we faced and not be stymied, at least for the time being, by a brain restricted by the size of the birth canal. And above all, because we now spent so much time developing outside the womb, we were enriched by the world and experiences around us, which made us smarter, more adaptable, and more individual. We benefited from the sights, smells, sounds, and relationships of early life, while our brains were still compliant—embryonic—enough to react to these experiences and learn from them. We were coming into the world less hard-wired, less DNA-driven, and more impressionable than any other creature.

Anthropologist W. H. Krogman put it this way: "This long-drawn-out growth period is distinctively human; it makes of man a learning, rather than a purely instinctive animal. Man is programmed to *learn* to behave, rather than to react to an imprinted determinative instinctual code."[10] In other words, it made us not simply capable of learning—a dog or a mouse is capable of learning—but adapted for learning, dealing with change, and changing still further in reaction to that change. Our ancestors now had brains that could better shape themselves to the world around them, rather than be restrained by the strict marching orders of their genes. And all of this pushed our ancestors more closely toward being human. But our youthfulness, and the helplessness that accompanied it, would have still further social implications.

. . .

Most mammals are born far better prepared to handle the challenges of life than we are. At birth a wildebeest is up and running with the herd within minutes. But the care that our predecessors' newborns required, even two million years ago, was considerably more complicated.

We already know hominid mothers needed help with the birth itself, but once the baby arrived, both mother and child also would have needed additional, serious support just to keep up with the daily migrations of the troop. Attending to her own needs, protecting her baby from predators, and simply managing to keep herself alive would have been a monumental challenge for any australopithecine mother. Not only that, if the birth rate was rising, she would have faced taking care of more than one child at a time. This had to have dramatically shifted the social and sexual dynamics within every troop. Finding mates—truly reliable, helpful mates—would have quickly become a matter of life and death. Maybe new mothers were able to count on the support of other females in the troop (this often happens with chimps), but only up to a point. Life was short and childbearing years few. There would not have been many females available who weren't already busy with their own offspring.

Sooner or later the most consistent help would have had to come from fathers, not midwives and cousins. This meant females needed to find mates with more to offer than broad shoulders, brains, stamina, and strength. They also had to bring primal versions of patience, honesty, attentiveness, and loyalty to the effort.

We should not underestimate the power of these changes, even though our understanding of how they may have played out is murky. Just as the heightened dangers of the savanna tightened the social fabric among the troop, the need to care for their increasingly helpless young inevitably drew both children and their parents into more emotionally intimate relationships.

We can see part of the evidence of this in the ways we relate with one another today. We are odd among the apes because we are the only ones that are even remotely monogamous. Male gorillas, for example, are polygenous. They have harems of females that they mate with, and jealously guard from competitors. Because the dominant male obviously has a strong set of genes, the system works fairly well, but only because male gorillas don't have to be strong helpmates. Female gorillas handle raising the young fine without their help.

For their part, chimps are polygamists. Both males and females mate with multiple partners whenever females are fertile. Recently scientists have found that a chimp's sperm has more mitochondria than human sperm. This actually gives the spermatozoons of different male chimps an increased ability to battle it out with one another in the womb to win the right to fertilize the egg of a female with whom the chimps have mated. (This is selection of the fittest at the most basic level.) Human semen doesn't have this capacity, and it very likely doesn't because there never was any overwhelming evolutionary need for it to develop. As a species we mostly stick with one mate (though there are obviously many exceptions), and the best theory as to why we do is that evolution favored savanna apes who mated and then worked together to raise and protect the helpless children they were bringing into a dangerous world.

This meant that somewhere in our hominid past something like a family unit began to evolve that also created increasingly complicated social relationships. Tangled strategies were required to figure out which mates would be the best caregivers, and the most reliable and loyal. Complex games of social chess had to have evolved to win the battles for one another's affection and fidelity. Obviously attraction to a strong set of genes continued to play a central role in selecting mates; this is why we still find big smiles that show off white teeth, and strong, healthy bodies and athletic prowess attractive in others. But now individual traits and personal behavior were also becoming increasingly important to making certain that the species and the troop survived.

The timing of all of these changes is hardly precise, but we can speculate that roughly three thousand millennia had passed since our ancestors and chimps had split off from their common ancestor. By this time multiple species of australopithecines had come and gone. *Homo habilis,* the first of our direct line, was now a central primate player on the savanna. It was edging increasingly toward humanlike intelligence, behavior, and relationships, all of which would both enable and require finer forms of communication. The first tiny fires of human culture were beginning to glimmer. New creatures were taking shape, and life on Africa's savannas was about to become more intriguing than ever. Because now that our predecessors had risen up and were standing on their own two feet, something else entirely new was in the process of evolving, and that trait would change them, if possible, even more profoundly than anything else already had.

II

Thumbs

Chapter 3

Mothers of Invention

Now, if some one man in a tribe, more sagacious than the others,
invented a new snare or weapon . . . the plainest self-interest, without
the assistance of much reasoning power, would prompt the other members
to imitate him; and all would thus profit. . . . If the invention were an
important one, the tribe would increase in number, spread and supplant
other tribes.
—*Charles Darwin,* The Descent of Man

L OOK AT YOUR HAND. Hold it up. Flex it. Bend it. Make it act like a
puppet. It's a remarkable piece of engineering. Never before have five
digits, fourteen joints, and twenty-seven bones come together in such an interesting and practical way. If you turn it, eight cubelike bones connected
by a matrix of tendons in your wrist and forearm enable you to rotate your
hand 180 degrees. This makes it possible to do things that animals in the natural world, even if they had the inclination, could never possibly carry off,
like swing a baseball bat, pour a glass of milk, play a Duke Ellington piano
solo, or paint a portrait.

The fingers of our hands actually have no muscles. They operate by remote control, like marionettes. A web of tendons, anchored in the palm,
midforearm, and as far north as the shoulder are the strings that make your
digits dance. The whole arrangement provides our hands with an unusually

wide range of motion. But the anatomical feature that renders your hand especially special is your first digit, its version of a big toe: your thumb.

One of the great beauties of our thumbs is their position. Whereas our feet forsook the thumblike position of their first digits and evolution straightened them into our big toes, our hands did not. In fact, they went in the opposite direction, building on their former status as feet specialized for climbing and grasping.

This is why our hands still look remarkably like a gorilla's foot, with our thumbs sitting down below the other four digits, positioned apart, as if they were reluctant to join the rest of the group. Not that this means, however, that thumbs in any way evolved to play second fiddle to the rest of our fingers.

Compared with the thumbs of other primates, ours have acrobatic ranges of motion. Chimp thumbs, for example, can't rotate in great swirling arcs like a human thumb, and that limits their ability to be the thing that all thumbs secretly long to be . . . opposable. I say "limits" because, contrary to popular belief, the thumbs of chimps and monkeys *are* opposable. They just aren't in the same peculiar way that ours are. What is different is that we can effortlessly swing our thumbs across the palms of our hands to meet our small and ring fingers, the fourth and fifth digits. Nothing like this exists anywhere else in nature. It's called the ulnar opposition, and this seemingly simple ability gives our hands the power to grasp and grip, turn and twist, manipulate and touch in ways foreign to other creatures. Because of this ability we can pick up and use a hammer or an ax, or turn a stick into a lethal club by cupping it in a position that extends the power of our arm, and, with it, the force of the blow it delivers. It is one thing to flail a stick horizontally for show, like a chimpanzee, another to grip it along the axis of your forearm and bring it down from on high with bone-crushing force.

Ulnar opposition also makes all of the difference between simply grasping a tree branch the way a chimps does when it swings through the forest, and precisely clasping minuscule objects with sweet and exact precision. When picking up something as tiny as a grain of rice, a chimpanzee has to squeeze it between its thumb and the flat of its index finger, like we hold a key or credit card. Such precise use of thumb and finger is a struggle for chimps because they don't have the musculature and nerve structure we do.

We can pick the same grain up using the very tip of our thumb and caress it in the closed circle of our finger, as if we were making the sign for "okay, perfect," which, in a sense, it is.

These abilities exist because we have developed specialized tendons linked to our thumbs. One, a flexor called the pollicis longus, runs from the thumb's knuckle all the way to the shoulder. Along with three other muscles, it lets us push and mash things as well as open our hands and spread our thumbs away from our palm, movements that come in handy when operating a joystick, typing on a keyboard, or thumbing in the numbers on a cell phone. But it is also very useful for gripping and manipulating sticks and stones, natural artifacts that our ancestors used to fashion the first tools into axes, spears, and small knives more than two million years ago.

It's not simply the speed and flexibility of our thumbs, fingers, and hands that make them special. It's also their extraordinary sensitivity. Crammed within every square inch of our digits are nine thousand hypersensitive egg-shaped, buds called Meissner's corpuscles, which lay just below the epidermis, our outermost layer of skin. Inside each bud lie coiled nerves that sense and snatch up the signals initiated by whatever we touch and send it to the brain for processing.[1] These same nerves are scattered among other particularly sensitive parts of our bodies—our tongues, the soles of our feet, our nipples, penis, clitoris—every erogenous zone. They're optimized for gathering the finest, most granular pieces of sensual information, and they are why our hands are, as Sir Charles Bell, put it, "so powerful, so free and yet so delicate."

Without this combination of dexterity and sensitivity, Michelangelo would never have been able to sculpt the face of his *Moses,* nor Leonardo paint *The Last Supper.* Horowitz could not even plunk out the most juvenile version of the *Emperor* Concerto, and Shakespeare would have been incapable of grasping a quill to pen a single word of the thousands he invented for the English language.[2]

The point here is a subtle one. The physical power and dexterity of our thumbs and hands make them central to our humanity. Their biological evolution literally changed our minds. They enabled us to better manipulate the world around us, and the manipulation of things then came to also mold our minds. This is what prompted novelist Robertson Davies to observe in his book *What's Bred in the Bone,* "the hand speaks to the brain as surely as the brain speaks to the hand." The wirings for creativity, for memory, for emotion, and above all (as we shall see) for language, exist largely because our thumbs came first, and in orchestrating our physical conversations with the world, laid the neural groundwork for the peculiarly human mind that would follow. Our thumbs are that defining. Without them, we wouldn't be human. We would be something else.

For the ancestral line of apes that led to us, there would have been a considerable evolutionary advantage in developing thumbs. As they gave up knuckle-walking and spent more time upright, their hands would have been freed to hold more, carry more, throw more, and eventually manipulate and make more. Had early savanna apes not begun walking upright, thumbs would never have evolved. And if they had not, we wouldn't have either.

. . .

Hands and thumbs go back in one form or another forty million years ago to prosimians, the line of mammals from which we evolved. But hands, at least the ones to which we have become accustomed, are relatively new. Based on the garbled messages the fossil record has so far provided, science's best guess is that they reached something like their current, thumb-opposable state a little more than two million years ago. By this time *Homo habilis* was emerging on the plains of Africa—brainier, faster, and more inventive than the other primates with which it was keeping company.

Exactly which line of australopithecines might have led to *Homo habilis* remains unresolved. But when this new genus surfaced, with its specialized thumbs, new and interesting events began to unfold. The first and most obvious change was toolmaking. Australopithecines like Lucy and the Taung child were tool *users* but not tool*makers*. Like chimps, they very likely put twigs, bone, grass, and rocks to work as weapons and various other kinds of primeval gadgets, mostly to help gather food.[3] But generally scientists agree they did not reshape them into anything sharper or more complex than nature originally intended, partly because they didn't have the dexterity, and partly because they didn't have the cerebral horsepower to consider the possibility. They were probably not very different from today's average chimp.

This dearth of toolmaking ancestors is one reason why the world was stunned in 1964 when primatologist John Napier, paleoanthropologist

Philip Tobias, and the dean of human evolutionary theory, Louis Leakey, wrote in *Nature* that they had found among the dust and rock of Olduvai Gorge in Tanganyika (now Tanzania) evidence of the first toolmaking creatures ever. This protohuman, Leakey, Napier, and Tobias told the world, was not an australopithecine. It had a larger brain than any ancestor found this far back in the fossil record. And it had a very humanlike hand. "The hand bones resemble those of *Homo sapiens sapiens* [modern humans]," they wrote, "in the presence of broad, stout, terminal phalanges on fingers and thumb . . ."

The scientific world dropped its collective jaw. A two-million-year-old hand that looked human was big news. But even bigger, perhaps, was the discovery of simple tools found in the same area and strata of rock as the creature's bones—flakes of sharp stone used for cutting and scraping.

Frankly, this surprised Leakey and his colleagues because the pieces of skull and jaw they had found indicated that the brain of *Homo habilis* was not quite as large as they would have expected in a true toolmaker—only about 680 cc (roughly half the size of the average human brain). Despite this, the scientists felt that the creature should be classified in the genus *Homo,* indisputably making it our direct ancestor. After all, if these creatures could make tools, they reasoned, whatever brains they had must have been enough to qualify them as one of us.

In the more than forty years that have passed since Leakey's discovery, anthropologists have debated exactly where *Homo habilis* fits into the family tree. Only four *habilis* discoveries have been made, so their pedigree has been difficult to resolve. But one issue that seems beyond debate is that *Homo habilis* had something that Lucy and other australopithecines before it didn't—elongated, fully opposable thumbs that are essentially identical to ours. As Napier put it, "The hand without a thumb is at worst nothing more than an animated fish-slice and at best a pair of forceps whose points don't meet properly. Without the thumb, the hand is put back 60 million years in evolutionary terms to a stage when the thumb had no independent movement and was just another digit. One cannot emphasize enough the importance of finger-thumb opposition for human emergence from a relatively undistinguished primate background."[4]

Habilis's thumb had evolved to the point where it made toolmaking possible. Because of its distinctly human shape and mechanics, *Homo habilis* could do something never before seen in the natural world: cup its strong palm and fingers around an odd and irregularly shaped chunk of flint rock, grab another smaller stone the way you might grasp a baseball (two fingers on top and the thumb below in a grip known as the "three-jawed chuck"), and repeatedly but precisely whack the larger stone.

Easy as this seems, no other primate can do it. Mary Marzke, who has made a career of understanding how our hands evolved and work, has noted that all of this was made possible "by a unique pattern of hand proportions and joint-and-muscle configurations that permit cupping of the hand and the formation of a wide variety of grips."[5]

In *Homo habilis,* evolution had shaped a hand that was the anatomical equivalent of a jack-of-all-trades. It could hold, twist, turn, push, and pull unlike anything that had come down the evolutionary pike. And this in turn made it capable of shaping itself in an unusually large number of grips and positions.

This was, to say the least, a good thing for *H. habilis* because he needed all the help he could get. By the time his kind had emerged, Africa's savannas were growing even less forested than they had been previously thanks to a new climatic heat wave. The clustered trees that remained, and the nuts and fruits they provided, were shrinking still further. But larger mammals, so-called megafauna, were continuing to evolve on the expanding grasslands, and they often fell prey to the savanna's big cats, leaving their carcasses

The "three-jawed chuck" is a grip unique to the human hand. It enables us to use the thumb, index, and middle fingers to grasp irregularly shaped objects, such as a stone, and use it as a tool, or transform a stick into a lethal club by cupping it in a position that extends the power of our arm, and, with it, the force of the blow it delivers. (Reprinted from "Evolutionary Development of the Human Thumb" by Mary Marzke. Used by permission of Mary Marzke.)

available to those animals that couldn't make the big kill themselves but didn't mind partaking of the leftovers. If *H. habilis* could become a resourceful scavenger, he might do pretty well.

Strong and nimble thumbs helped him accomplish that. With them, and the hands they made possible, he transformed chunks of stone into honed knives with exceptionally sharp edges and then put them to work butchering animals the size of hippopotamuses and elephants. These weren't a hunter's weapons; they were artificial versions of a jackal's jaws or a vulture's beak. Carrion tools. But they represented vitally important advances. At least that is what fossil finds in Olduvai Gorge have indicated.

To test just how well knives of this kind might actually have worked, Nicholas Toth and his archaeologist colleagues from Indiana University visited the same locations where *H. habilis* lived two thousand millennia ago. Once there they took into their hands the same flint rock out of which *H. habilis* had fashioned tools and created their own knives by carefully hammering a small chunk of rock against a larger "core" stone. With every blow of the "hammer," a razor-sharp shard of the core would fall to the ground.

East Africa, they found, was loaded with the raw materials used to fashion the first artificial tools. The stone knives were not difficult to create, if you had the thumbs for it, and they managed to produce exact replicas of the ancient tools again and again. Then came the hard part.

They took the sharp-edged stones into the savanna, where on two different occasions they located the carcasses of elephants that had recently died of natural causes. And then, like members of a small troop of *Homo habilis,* they set to work flaying and carving the animals up. Toth and Kathy D. Schick described the experience in their book *Making Silent Stones Speak: Human Evolution and the Dawn of Technology*:

"Somewhat daunted, we approached our task equipped with simple lava and flint flakes and cores, which looked more and more paltry as we got closer to the impressive body. Initially, the sight of a twelve-thousand-pound animal carcass the size of a Winnebago can be quite intimidating—where do you start? We had never seen a field manual on pachyderm butchery, and they aren't like smaller animals: you cannot move the body around (for instance, flip it over to get a better vantage) without heavy power machinery. You have to play the carcass where it lies . . .

"Despite the success of our tools in dozens of other butcheries, we were not really sure they were up to this task. We were amazed, however, as a small lava flake sliced through the steel gray skin, about one inch thick, exposing enormous quantities of rich, red elephant meat inside. After breaching this critical barrier, removing flesh proved to be reasonably simple, although the enormous bones and muscles of these animals have very tough, thick tendons and ligaments, another challenge met successfully by our stone tools."[6]

. . .

These tools and abilities bestowed *H. habilis* with an evolutionary edge no other animal had ever enjoyed. At the very time *H. habilis* was carving carrion, other bipedal primates, such as *Paranthropus boisei*, were living nearby taking another approach. *P. boisei* ate tubers, grubs, berries, and nuts, not meat. But with his stone knife, *H. habilis,* who already had a larger brain, was now able to dine on deinotherium or hippopotamus carcass, which in turn provided increased health and the raw protein needed for growing still larger brains.

Not that he was a master hunter. At four feet and a hundred pounds, he was far from fearsome, but thanks to his tools he could piggyback on the strength, speed, and ferocity of other animals and supply himself with food

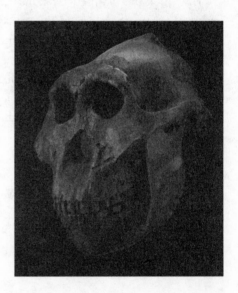

An adult *Paranthropus boisei* skull.
(Used by permission of the Smithsonian Institution.)

that eluded his cousin primates. In time *H. habilis*'s handmade tools put more evolutionary distance between himself and the australopithecines roaming the savanna. He was a technological animal with all of the advantages technology affords. While the fossil record indicates that *Paranthropus boisei* and its relatives actually escaped extinction longer than *Homo habilis*, eventually their line led nowhere. *H. habilis*, however, evolved into other, increasingly intelligent species of toolmakers, such as *Homo erectus* and *Homo ergaster*, lines that eventually led directly to us.

. . .

While *H. habilis*'s tools enhanced his survival and accelerated brain growth, they set even more far-reaching events in motion. The thumbs that made tools possible were also shaping a new kind of mind. Leakey and Napier eventually argued that it wasn't the tools alone that *H. habilis* fashioned that separated him from the rest of the primate pack, it was his mind. Or more accurately, it was a brain capable of conceiving and making tools that truly distinguished him.

It was Raymond Dart who suggested to Leakey and his colleagues that they name the new, toolmaking creature they had discovered *Homo habilis*. Most times this is translated as "handy man," but as the team pointed out in their paper, *habilis* also means "able, mentally skillful." That insight may have turned out to have been more accurate than even Leakey and his codiscoverers suspected, because as scientists have realized since, mental skill and manual dexterity go very literally hand-in-hand.[7] Or put another way, the physical world in which our ancestors evolved shaped the mental world in which we live today. The two of them cannot be separated.

. . .

More than twenty-three hundred years ago, the great orators of ancient Greece (and later Rome) employed a terrifically creative technique for remembering long speeches and poems. They called it *lopoi*, the Greek word for places. (The equivalent Latin word, which we still use today, is *loci*.) Mnemonic devices were a necessity in those days. Paper and pen were still rare, and it wasn't a simple matter to jot thoughts and passages down like we do. Orators like Demosthenes and Cicero would sometimes speak for hours, enthralling their audiences and devastating their opponents in debate while they used nothing more than loci to track the threads of their logic.

They did this by imagining a physical space they were familiar with. Think of walking up to your home. There is your front porch, your door, the hallway and living room. Imagine you want to remember a grocery list. You simply picture yourself walking up to your house and associating the things you want to buy with each location: milk on the porch, apples at the door, bread on the floor of the entry. Once you have linked an item with a specific location in your mind, all you have to do is walk again through your imaginary house, and each reminder is there waiting in the right order. With a speech it's a little more complicated than this (the "items" are more abstract), but the idea is the same: associate what you want to remember with moving through familiar physical spaces.

There is no obvious reason why this visualization technique should work any better than simply memorizing a list of concepts (or groceries) you want to recall, but it does. And it does because brains evolved to map the physical world long before they evolved to handle abstract thought. We live in three-dimensional physical space. We move forward and backward, left and right, up and down. On the most basic level our gray matter relates to the world in these very physical terms. Even the simplest bacteria and the smallest fish "understand" the world this way. If they didn't, they would be motionless and paralyzed—incapable of escaping a predator and powerless to pursue food when they sensed it. Being alive requires understanding space and moving through it.

We tend to assume that the thinking that buttresses high-minded and complex abilities such as language, philosophy, strategy, reflection, invention, and creativity are not truly connected with the physical world. But there is mounting (some argue irrefutable) evidence that because our brains evolved to move us through physical space, that evolution has deeply shaped the way we think about everything. Linguist George Lakoff and philosopher Mark Johnson have pointed out that as loaded as our mental lives are with intangible concepts such as importance, similarity, difficulty, desire, intimacy, and ambition, we actually think about them in very concrete terms. We "see" what someone means. We "grasp" an insight. If a concept eludes us, it goes "over" our heads. We "crush" opponents; "fall" in love; "kick" ideas around; feel "squeezed" when we are under pressure; and "hold" someone in high esteem. We even express emotions in terms of distance or height. We are "close" to our friends, "distant" when angry. We feel "down" or "up." If something is important, it is "big." If a movie or a book is bad, it "stinks." Even something as abstract as time is conceived and expressed in physical terms. The past is "behind" us and the future lies "ahead."

Metaphors like these are pervasive in every language and throughout human thought, whether you hail from Mongolia or Tierra del Fuego.[8] And they are wired into the human brain as early as infancy and toddlerhood. Johnson calls this process "conflation." Babies, he says, are not mentally capable of fully separating the experience of one thing that is often associated with something else in their lives. For example, the affection an infant experiences is usually associated with the warmth and security of being physically held, so she "conflates" the two experiences—being close to someone physically equates to the security that closeness creates. Later in life, of course, we learn that affection and physical warmth are not the same thing, but because they were in our infant experience, we continue to link them conceptually.

Other experiments indicate that when we think of the word "fall," for example, we may experience feelings of fear and failure that we associate with falling because neural connections in our brain have physically connected them at the synaptic level.[9] This means our talk of "warm smiles" and "close friends" may be neurologically, as well as conceptually, harnessed together.

This very likely explains why we find it easier to remember ideas when we associate them with a physical activity, such as walking through our house. But with the arrival of hands, especially thumb-opposed hands, our brains developed an even more physically precise perception of the world because now rather than passively reacting to our environment, thumbs made it possible to intentionally grasp and manipulate it in ways nature had never seen before. That would link two shattering events in human evolution that most of us might not assume are connected: toolmaking and language.

. . .

Patricia Greenfield, a developmental psychologist at UCLA, has found that there are remarkable connections between the way children use their hands to control objects, and the ways in which we all organize symbols in our minds before expressing them in words.

In one test Greenfield asked children of three different ages—six, seven, and eleven years old—to solve a puzzle.[10] Twenty sticks were laid on a table and positioned so they formed a series of connected boxes. Each child was then given an identical set of sticks and asked to re-create the same pattern of boxes. The way each group approached the problem provided some surprising clues into the way our brains organize thought.

The six-year-olds all had a characteristic way they went about tackling the problem. They laid down a stick, and then used the next stick to connect to the last one. They never created a separate box that wasn't linked to the sticks they had already laid down. In fact, they always linked the stick in hand with the last one they put on the table. Basically, they were feeling their way through the problem until they painstakingly reconfigured their sticks to look like the pattern that had already been laid out before them. They really were incapable of doing it any other way.

Seven-year-olds dealt with the challenge a little differently. They didn't always link the last stick to the previous one. They sometimes created a separate, disconnected box here and then another there. Then they linked separated boxes with other sticks until the problem was solved. These children weren't feeling their way through the puzzle. They were far more creative and confident than children only a year younger, according to Greenfield. Rather than slavishly imitating the pattern stick by stick, they were now generalizing the whole pattern, which they then went on to build in separate chunks that they eventually connected.

Eleven-year-olds made a quantum leap. In fact, when they were presented with this puzzle they treated it something like the way a virtuoso pianist might treat playing "Twinkle, Twinkle, Little Star." They seemed to stow the whole pattern in their minds at a glance, where they mentally held it while they considered other aspects of the problem. (This is a unique human ability called "working memory," which allows us to put mental constructs aside while we tackle something else without losing track of the original concept). In any case, the eleven-year-olds didn't see the puzzle as a stick-by-stick problem, or even a box-by-box problem. They played with the whole pattern, re-creating it in all kinds of ways, playfully; a stick here, a stick there, boxes connected by boxes, whatever they could imagine. For them it was all simple.

Greenfield believes that children at these ages approached her problem so differently because specific parts of the brain involved in organizing objects have to physically link before they can solve the puzzle in increasingly sophisticated ways. She also believes that the ways in which children physically re-create the puzzle—the way they choose to connect the sticks—parallel the ways in which they organize their thinking and their language.

Consider, for example, the approach we take at an early age to stringing together words. Very early on we manage it well enough with basic information, like "Me want bottle." But later, ideas become more complex and words are arranged in more complicated ways to communicate more intricate concepts. "May I please have the bottle?" (Well, usually it's "*Can* I please have the bottle?") Or, "May I have the bottle of milk over there on the kitchen counter?" The point is that we figure out where to place words and ideas in the patterns of our sentences and thought not unlike the ways in which we figure out how to place objects in a puzzle. Words serve the same function in language as objects do in Greenfield's stick-box experiment. If you can put the sticks (words) together in the right configuration (syntax), you solve the puzzle and construct something that makes sense. The language equivalent of this is first the conception or acquisition of an idea, and then a sentence that expresses it. In this way, words and concepts are like virtual, imaginary objects we move around in our minds, not unlike objects we move in the physical world.

Greenfield's research has led her to theorize that the advances children make in her experiments parallel the mental evolution of our predecessors. But what puzzles would our ancestors have been trying to solve two million years ago, and how would they have shaped the brain we have today?

· · ·

Because we have only two hands, rather than, say, eight tentacles, like an octopus, we manipulate objects in an ordered sequence, not all at once. That means to consciously do "A" before "B" and "B" before "C," we have to focus. You don't absentmindedly build a bow, or shape an arrow, or design a steam engine. It requires intention and concentration. Anyone who has struggled with assembling furniture at home knows that if B does not follow after A and C upon B, things have a way of falling apart.

If scientists such as Lakoff, Johnson, and Greenfield are right, we manipulate thoughts the way we do because our hands once learned to shape sticks, stones, and animal skins into tools. Nouns became the equivalents of objects, verbs represented actions, and we (or our hands) took on the role of a sentence's subject.

To ancestors like Handy Man, the physical grammar of cracking open a femur to eat the marrow inside might have gone something like, "Hit bone (with) stone." He might not have had any words—any mental symbols—to attach to these objects or actions, but the pattern of using one thing to affect another would have been part of his physical experience. There was no way around it. If you pick up a stone to strike a bone, certain actions must unfold in a certain sequence for the whole business to work out. The brain must consciously conceive and act on that sequence, or the bone and stone will forever sit there, and never the twain shall meet. And any ape that spends his day gazing at a rock and bone, doing nothing, will never eat an ounce of marrow, and certainly won't live long enough to pass his genes along. Animals like these, as scientists like to say, "get selected out."

The unavoidable conclusion here is that toolmaking not only resulted in tools, but also in the reconfiguration of our brains so they comprehended the world on the same terms as our toolmaking hands interacted with it. The physical conversation our marionette fingers were having with the objects around us was shaping the way our brain organized and thought about everything. The hand speaks to the brain as surely as the brain speaks to the hand.[11] Art, or at least craft, was beginning to imitate life, and the rudiments of language and complex human thought were sprouting from the sense-able, concrete sequences of that life.

. . .

In 1996, Vittorio Gallese, Giacomo Rizzolatti, and their colleagues at the University of Parma in Italy inadvertently discovered the strange and mysterious ways in which evolution works. They were recording signals transmitted from neurons in an area of the brains of macaque monkeys called the F_5 region. This is a specific sector of the frontal lobes that sits among a larger area of the brain that deals with making and anticipating movements called, fittingly, the premotor cortex.[12]

The scientific team already knew that F_5 neurons fired when monkeys performed specific goal-oriented tasks with their hands or mouths—picking up a peanut and then holding it, for example. But for this series of tests they wanted to see if the F_5 neurons acted any differently when the objects themselves were different. Did it matter, they wondered, if a monkey was picking up a peanut rather than a slice of apple?

It was while they were performing this routine experiment that they noticed something odd. When a macaque watched a researcher's hand pick up an object and bring it close to his mouth, the sensors connected to the monkey's brain indicated that neurons in its F_5 region were firing. They didn't activate when the monkeys simply saw the objects sitting there, only—and this was what was so unusual—when the monkey watched researchers pick them up, or when the monkeys themselves picked them up.

The implications of this are enormous. If the same neurons were firing in the monkeys' brains when they watched the action, it meant they were playing out what they were seeing before them inside their own brain—their mind's eye—just as if they were doing it themselves. They were mentally "mirroring" the physical action. You could also say that in a rudimentary way they were imagining they were *doing* the action; reliving, neuron by firing neuron, the experiences of others—in effect, putting themselves in the shoes of the researchers they were watching. They were experiencing a form of empathy that itself required a kind of imagination.

The ignition of F_5 neurons made these seemingly simple gestures and maneuvers a form of communication far more powerful than any hoot, grunt, or howl. After all, if the monkey was mentally picturing the actions of the researcher, it was also quite possibly remembering and learning it. Monkey see, monkey do.

If you look hard, you can catch glimpses of early conscious communication on all sides of this. Imagine two habiline creatures—a parent and a child—sitting in their small, lakeside camp two million years ago, smoke billowing from the enormous volcanoes at their backs. They have roughly twice the neuronal wetware of the average chimp today (and certainly more than a macaque monkey), so their intelligence is far from trivial. On the other hand, they still can't speak, so their ability to share what is on their minds is limited, even though they undoubtedly have far more to communicate than any of the other animals around them.

Now imagine the parent is making a simple tool, like those that Nicholas Toth and his colleagues experimented with. The child watches intently. Simply by observing, the same neurons—her mirror neurons—are firing in her head that are firing in her parent's. And so when she attempts to repeat the action she has been watching, she can call upon those fired neurons to guide her hands to do something she has never actually done before but *has* imagined doing.

For his part, when the parent strikes flint against the rock, he is silently talking to the watching child. He is saying, with his hands, "This is how you make this thing. You hold this large rock like this and strike it with this small rock just so." You can see him holding up the sharp sliver of flint that the blow has created. "See, now you have a knife." And then next, he may carve the skin off a carcass, taking the "conversation" in a new direction.[13]

The entire time the child is "listening." Neither parent nor daughter have any language; not a single word they can exchange, not even a concept of words, only the looks on their faces, the expressions in their eyes, the gestures they make with their hands as they manipulate and exchange the rocks and flint. But a lot of information is traveling back and forth between their two minds. In a very real sense they are conversing.

This apparent connection between conversation and manipulation is more than metaphorical. More recent research, built on Gallese's and Rizzolatti's original discovery, has revealed that the F_5 region in macaque monkeys is an analog for areas in our own brain essential for generating human language and speech (not necessarily the same thing, as we shall see). We know this partly because a few years after the discovery of mirror neurons, Rizzolatti and another researcher, Scott Grafton, found that when humans watch someone handle objects, a region of the brain called the superior temporal sulcus, which sits directly behind the left temple, activates and mirrors what they see. This surprised scientists because they had long thought that this part of the brain existed primarily to send the signals to Broca's area that generate speech. Now it appeared Broca's area was handling other jobs as well, or deeper ones. It not only sent signals to the muscles that generated speech, it sent the signals to hands and arms that enabled the precise manipulation of objects.[14]

Rizzolatti thinks this fusion of objects and imagination, gestures and words provides a glimpse into the genesis of language. Mirror neurons might be the primal wetware that enabled our ancestors to transform the common ground of doing and making into the earliest forms of conscious communication. F_5, or something like it, might very well have been the bud from which Broca's area—a cornerstone of human language—blossomed.

The Insights of Dr. Broca

How we actually generate language is a mystery, but we know that we can't do it if a part of the brain known as Broca's area, named for the brilliant French doctor and anatomist Pierre Paul Broca, who discovered it, doesn't function properly. Broca first located this part of the brain when he performed an autopsy in 1861 on a patient, known as Tan, who had died from gangrene. The man was known as Tan because when he tried to speak all he seemed capable of saying again and again was the word "tan." This affliction became known as Broca's aphasia, and the autopsy revealed that there had been damage to specific sections of the inferior frontal gyrus in the left frontal lobe of the brain (roughly near the left temple). Subsequent studies Broca and others performed confirmed that in most people (left-handers usually being the exception) this is the area of the brain that somehow takes the symbols our minds create when we want to communicate, attaches sounds to them, and then coordinates sending the signals to all of the muscles needed to make the precise sounds we call speech (or in the case of those who can't speak, make the hand signals needed to communicate).

Brain scanning technology has confirmed Broca's findings. These areas of the brain "light up" when we generate speech. Broca's area is connected to Wernicke's area by a neural pathway called the arcuate fasciculus, and using these two sectors of the brain, we handle most of the generation and understanding of the spoken (or signed) word. Because Broca's area is so closely located next to areas of the brain associated with mirror neurons and those sectors that control both facial muscles and hand coordination, it may help explain how toolmaking, gestures, and speech are connected.

With mirror neurons, something entirely new had entered the world: a far more effective and speedy method for pooling knowledge and passing it around than the old genetic way. Ideas could now be shared between minds! And that sort of knowledge-pooling, as Darwin observed, would have seriously improved the chances of a troop's, a family's, or an individual's survival. As he put it, "the plainest self-interest, without the assistance of much reasoning power, would prompt the other members [of a tribe] to imitate him; and all would thus profit. . . . If the invention were an important one, the tribe would increase in number, spread and supplant other tribes."[15]

This means that two astounding advances were unfolding during *Homo habilis*'s brief stay on Earth. First, entirely new knowledge was being intentionally generated out of the brain of a single creature. Toolmaking marked the birth of invention. Second, knowledge could now be duplicated and relocated to other minds; it was no longer doomed to die with the brain that conceived it. Just as the evolution of DNA made it possible for a gene to be copied and shared from one generation to the next, mirror neurons, and the new behaviors they made possible, meant that an idea—a "meme," as Richard Dawkins has put it—could be copied and passed along from one mind to the next.[16] Conscious communication had emerged, even if only in an embryonic form, and in its wake everything from gossip to oratory, mathematics to the laws of Hammurabi, stand-up comedy to the computer code that sends probes to the moons of Saturn would follow. We were building the scaffolding for true human behavior, relationships, and, ultimately, that most monumental of all human inventions: culture.

But how would our ancestors even begin to cross the chasm that yawned between the first flint knives and the great edifices of human endeavor we have erected since?

62

Chapter 4

Homo hallucinator—the Dream Animal

*Man is a singular creature. He has a set of gifts which make him unique
among the animals: so that, unlike them, he is not a figure in the
landscape—he is a shaper of the landscape.*
—Jacob Bronowski

We inhabit a language rather than a country.
—Emile M. Cioran

Colorless green ideas sleep furiously.
—Noam Chomsky

HUMAN CULTURE REQUIRES MASSIVE mind-sharing. Inventions
such as money, trade, government, religion, literature, and agriculture
are cooperative ventures that demand intricate bridges, trestles, and aque-
ducts of communication, and for those we need language. It shapes us as
much as we shape it.

Philosophers, linguists, and anthropologists have been debating the origins of language for centuries, and still the arguments tumble on. Some theories date back to the nineteenth century, and today many aren't taken as
seriously as they once were. There is the Bow-Wow or onomatopoeic theory,
which holds that speech arose when our ancestors imitated environmental

sounds—the oink of a boar or the sounds of the wind. Words such as "crash," "whoosh," and "bang," which sound like the actions they describe, are good examples.

Then there is the Pooh-Pooh theory, the idea that language arose out of instinctive cries such as "ouch "or "oh" or "yikes!" The Ding-Dong theory suggests that our ancestors reacted to the world around them by spontaneously producing sounds we associated with a person or a thing. Mama, for example, might have evolved out of the "mmmm" sound that goes along with nursing.

There are other theories. Ethologist and linguist E. H. Sturtevant has wondered if human language developed because humans found a selective advantage in being able to deceive others. Exclamations such as a yelp or a moan can involuntarily reveal your true mental state, so, according to Sturtevant, humans learned to fake them to deceive others for selfish advantage. There may be something to that, but it doesn't really solve the problems of the mechanics or cerebral wiring needed for speech.[1]

Psychologist Peter MacNeilage at the University of Texas at Austin theorizes that Broca's area evolved in humans as the brain center that produces speech because that part of the brain also handles sucking, biting, and swallowing. Maybe, he argues, these functions helped to frame and separate the sounds we make that eventually became words.

Darwin at least partly subscribed to the theory that language evolved from the sudden, unconscious noises our ancestors made (the Pooh-Pooh theory). "[Anyone] fully convinced, as I am, that man is descended from some lower animal," he wrote, "is almost forced to believe a priori that articulate language has developed from inarticulate cries."

These theories basically hold that somehow we began to make noises that we came to attach to actions and things in the world around us, and then began to string them together into simple pidgins, or protolanguages. Proponents agree that these strings of sounds may not have worked very well, but they were more effective than body language and limited facial expressions. Whatever the case, from these protolanguages, the theories go, modern language developed.

But what these models don't explain is how our ancestors began putting words in an order that wasn't simply random. Pidgins, for example, pile words together without much regard to their order, which severely reduces their effectiveness. Take linguist Derek Bickerton's example of this Hawaiian-Japanese-English pidgin construction: "*aena tu macha churen, samawl churen, haus mani pei* [translation: and too much children, small children, house money pay]."[2] You get a sense of the meaning of this sentence from the words in it, and it is better than nothing, but it is a long way from anything as full-blooded and profound as, say, Shakespeare's "Tomorrow, and tomorrow, and tomorrow creeps in this petty pace from day to day to the last syllable of recorded time. And all our yesterdays have lighted fools the way to dusty death"—two sentences bristling with metaphor, emotion, insight, and organization.

The order of words in a language is at the heart of its syntax, and syntax is the bedrock of grammar, the organizational rules that define and differentiate the underlying structure of every language. Vocabulary may supply the bricks and the mortar, but without syntax and grammar there would be no such thing as language as we know it—they supply its shape and design, its studs, struts, foundations, and supports. How could all of this come together, and how did a brain that could both create and comprehend it emerge?

· · ·

The problem for linguists and anthropologists when they set off to decipher these mysteries is that their resources are limited. No concrete examples of early languages exist. There are no ossified grammars or words, and while we have the skulls of our ancestors to inspect, they don't yield many hints about the convoluted structures of the living brains they once housed. There are clues here and there, but they provide little more than a smoking gun, or sometimes just smoke. But there are some other sources of information that scientists are managing to turn up—among them, the *non*verbal ways in which we communicated *before* true language emerged. These have primeval roots that find their way back to the most basic forms of animal communication.

Some mammals, when they are threatened or square off to fight, raise the fur on their bodies so they look larger and more menacing. When the hair on the back of your neck rises, or you get goose bumps on your forearms walking by a graveyard, that's a legacy of the same primordial behavior. You are scared, so your first reaction, without a thought, is to raise the fur on your body so you will look fierce, even though you don't have any fur left to raise. Birds fan their feathers, plump their plumage, or break into exotic dances and resplendent arias to win the attention of the opposite sex. When a wolf bears its teeth and growls, the message is clear: Get out, or prepare for the worst. And what says more than the wagging tail of a dog?

There are other nonverbal forms of communication. To express dominance in a group, lions or gorillas will charge the challenger, and most times the challenger rolls over on its back, and shows its belly to concede defeat. Horses have a biting and kicking order, just as chickens have a pecking order. The unfortunate horses and chickens that find themselves at the low end of the totem pole accept the abuse. It's a form of communication that lets every animal in the group know where they stand.

This unconscious language that bodies speak is as wired into the human race as it is into other species. When physically attacked, our first instinct is to put our hands and arms around our heads, duck, crouch, basically withdraw our bodies from the attacker as much as we possibly can, something scientists call "tactical withdrawal." It is a form of flight when we can't actually flee. We have developed other subtle forms of tactical withdrawal from verbal attacks and confrontations, too. We hang our heads, wince, or shrug, for example, all motions that send subliminal signals of a desire to be somewhere else.[3]

Evolutionary psychologists theorize that body language evolved as a visceral, external communication of the inner state of the creatures displaying it. It uses the usually unintentional vocabulary of our limbs and muscles to reveal what is on our minds or in our hearts. That is one reason why when we speak with someone face-to-face, our bodies are often holding a parallel conversation with them that can be substantially different from the messages contained in the strings of words we are exchanging. We have all finished a conversation or meeting feeling down or confused or unusually good, not because of anything that was actually said, but because we have subliminally downloaded messages of sadness or deception or joy expressed in the bodies and faces of the people we just met with.

Darwin wrote about the nonverbal ways both animals and humans communicate in his 1872 book *The Expression of the Emotions in Man and Animals*. He speculated on everything from what it meant when a cat pins its ears back to the way we turn down the corners of our mouths when we are sad. Seventy-five years after Darwin explored those ideas, anthropologist Edward T. Hall and psychologist Paul Ekman, among others, introduced the world to kinesics, the science of body language, and began to more closely explore the meanings that lay hidden behind crossed legs, licked lips, and raised eyebrows. One study even showed that when we see something we like, not only do our eyes widen (presumably to get a better look), but our pupils do, too, something good poker players keep in mind.

Our bodies speak at this primal level because the body's language is ancient. Its messages travel along paleocircuits, nerve pathways that were set in place millions of years before the brain had assembled the hardware used for speech and conscious thought. Some body language, such as shrugging, traces its origin to the motor pathways that existed in the spinal cords of jawed fish that first began to swim Earth's seas during the Silurian period 420 million years ago. When we push our palms down or straighten our backs and raise our heads in a conversation, we are recapitulating behavior traced to reptiles that developed a predisposed, genetic ability to look larger and fiercer by temporarily rising up on their hind legs.

Still other studies have shown that lower rates of head nodding during conversation can be associated with deceitful communication. In fact, there are potentially hundreds of nonverbal cues related to deception, from increased blinking to movements such as rubbing our noses or necks or eyes as we speak. If we are suddenly caught off guard in a conversation or find ourselves in a personally uncomfortable position, we might gulp or swallow hard, and bob our Adam's apple like a yo-yo.

Even the way we stand in relation to those around us says a lot about whether we feel uncomfortable, threatened, open, or respectful. Our bodies tend to square up with the people we admire.[4]

Scientists believe that most body language is rooted in the irreducible basics of a species' survival—fight, flight, submission, courtship, even disgust with rotten or poisonous foods. Eventually these ancient reflexes evolved into reactions that revealed increasingly complex inner states, such as fear, anger, or personal revulsion. And then those became important forms of communication; ways of saying, "I submit. I'm scared. I'm happy. I want to mate!" They laid the first foundations of communication, which grew more sophisticated as creatures became more intelligent. (The body language of a dog, for example, is more complicated and more telling than the body language of a gecko.) In time, especially with primates, body language engaged new parts of the anatomy to express information and emotion—faces, for example.

. . .

The faces of most four-legged mammals act not as communicators but as weapons. They are the end point of a projectile outfitted with snouts for smelling, eyes for tracking danger, and teeth for attack or defense. But when our ancestors stood upright, our faces no longer sat at the ends of our bodies, and that changed their appearance and function.[5] The fossils of every new version of upright savanna ape that scientists have examined have so far revealed that with the passage of time, foreheads rose to make room for rapidly growing brains; flat, fleshy noses increasingly protruded; and snoutlike jaws eventually morphed into the squarer chins and the flatter cheeks that characterize our human looks.

We already know that our ancestors were born at progressively earlier stages in their development, and increasingly carried that high-browed, youthful look further into adulthood. The line from their foreheads to their chins steadily grew straighter. Their increasing use of tools and weapons for hunting, meat-eating, and battle reduced the need for the boulder-shaped molars used to grind leaves and nuts, and fanglike incisors used for biting both competitors within the troop, and predators outside of it. This reduced the size of their jaws, and along with their altered foreheads, moved their eyes to the center of their faces, where they sat above the broadening planes of their cheeks. All of these rearrangements transformed their expressions into better placards for displaying their thoughts and emotions.

Over time savanna apes developed new ways to put those placards to work. Of the thousands of species of mammals on Earth, the human species owns the most expressive face. It has forty-four muscles, twenty-two on each side, about twice as many as a chimpanzee. These muscles not only adhere to bone, but also to one another and to the skin above them, enabling us to arch our eyebrows expressively or turn on a beaming smile. Our faces speak volumes of subtle emotion with just the tiniest frown, a quick wink, a pout, or a piercing, skeptical stare.[6]

Face-to-Face

Psychologists agree that we use our faces to communicate six primary emotions: happiness, sadness, fear, anger, disgust, and surprise. They disagree a bit on three others: contempt, shame, and startle. Startle, you would think, is closely connected to surprise, but some researchers feel it is unique, closer to a visceral reaction than to an emotional one. Contempt may be related to disgust, but some psychologists argue that the behaviors that elicit contempt are slightly different, and the facial movements that express it are more asymmetrical than disgust, which has its origins in avoiding dangerous food. Maybe one evolved from the other. There's no way to know. The same is true of shame, which also may be related to disgust—in this case, disgust with ourselves. Whether these facial expressions were also used by immediate ancestors such as *Homo erectus* we may never know.

Whatever the case, these reactions are so primal and deeply engrained that they are very difficult to hide and extremely tough to fake. And because of how focused our facial expressions are—because they are front and center in this little area, lighting up like neon signs—they have enormous impact. In conversation facial expressions seem to buttress what we are saying more effectively than a slouch, a rubbed neck, a turned shoulder, and other body language. When someone is surprised by something you say, you recognize the surprise in her face immediately; there is no mistaking the meaning. The same with a smile or a frown.

The interesting thing about facial expressions is that they seem to arise without a lot of deep conscious thought, but they are more intentional than body language. We don't usually think, "I believe I'll frown at that." It simply happens, before we've had a chance to think. On the other hand, we can sometimes decide to consciously smile for any number of reasons—because we are happy, because we are uncomfortable, because we are covering up another feeling. In the case of speech, we almost always think about what we intend to say before we actually say it.

In addition to the specific expressions that have evolved to symbolize clear states of mind, we use a second category, called emblems, which reveal culturally specific symbolic communications, such as a wink. Winks mean something in the United States but nothing at all on Easter Island. We also use manipulators like lip biting; illustrators (raised eyebrows); and regulators, which guide and direct conversation, actions such as nods or head turns, smiles, or a wrinkled brow.[7]

Facial expressions that communicate primary emotions such as happiness and sadness might straddle the worlds of the unconscious and the premeditated. They lie in the no-man's-land between a purely visceral, physical reaction such as a scream and the methodically conscious communication of a lawyer's summation before a jury.

Neurologists have found that such facial expressions as grimacing or setting our jaws trace their roots to the primal behavior of teeth-bearing. And the wide, white eyes of fear and horror, or the pursed lips of stubbornness and anger, find their origins in the evolution of mammalian midbrain nerve bundles in the cingulate gyrus facial circuit, which run along winding pathways from a brain area called the anterior cingulate cortex. They then pass directly through the hippocampus, the amygdala, and the hypothalamus—three key emotional centers in the brain—to link to the large cranial and facial nerves that control the larynx and muscles that make sound and move our lips.

It is almost as if, still bereft of speech, evolution had found a way to step beyond the more vague language of our ancestors' bodies to provide them with an increasingly precise vocabulary of movements focused in the one place we could never miss when dealing with one another now that we stood upright. Our faces became an emotional billboard, a miniature, more subtle, and more focused way to speak with our bodies.

Though we can tell from the skull fragments of *Homo habilis* that he looked far more simian than human, his forehead *was* higher and his jaw smaller than the australopithecines before him. His face was probably hairless in the way his chimp cousins were. (Although it is not immediately obvious, chimps *are* bare-faced. It's just that their low brows, muzzles, and smaller cheeks reduce the hairless area.) And, as *Homo habilis* moved across the plains of Africa, fashioning his simple, sharp tools, scavenging and gathering, competing and cooperating, this new face would have more easily expressed the increasingly complex emotions behind it.

However, it is in *Homo habilis*'s evolutionary successor that we see a creature where for the first time a primate looked more like us than an ape. That was an important turning point in more ways than one.

. . .

About one and a half million years ago, an adolescent boy died making his way along the shores of Lake Turkana in what is today Kenya, Africa. We don't know why or how; fever perhaps, possibly a predator ran him down, maybe he had somehow become separated from the troop and couldn't find his way back. But when paleoanthropologists Alan Walker and Richard Leakey uncovered his remains in 1984, they realized they had discovered a new kind of creature. They called him Turkana Boy and named the species that he represented *Homo erectus* (the erect human).

Since the discovery of Turkana Boy, other fossil finds have revealed *H. erectus* as the ultimate savanna primate. He was bigger, stronger, faster, and more mobile than *Homo habilis,* built for speed and distance—a hunter, not a scavenger, that migrated far beyond Africa.[8]

From the neck down, *H. erectus* looked remarkably similar to modern humans, except that he was even more genetically optimized for running than we are. His rib cage was virtually identical to ours. His hips were actually more narrow and his femur and tibia, unlike the savanna apes that preceded him, proportioned exactly as ours.[9] And he was tall. Leakey and Walker found an almost complete skeleton, and though Turkana Boy himself was only five feet, three inches tall, his bone structure and dimensions indicated he was an adolescent. Had he survived, Walker and Leakey estimated he would likely have grown to six feet in adulthood. When he walked or ran, his stride would have been extraordinarily graceful and efficient, an ability that would have served him well because, unlike *H. habilis,* he was a big-game hunter, if the fossil tools that have been discovered with other *H. erectus* remnants are any indications. Chief among these was the hand ax, a tool that begins to show up in the fossil record about 1.4 million years ago.

The hand ax was a Stone Age version of the Swiss Army Knife, and considerably more sturdy and refined than the small cutting tools *H. habilis* used. It looked something like a large arrowhead with a pointed tip, sharp edges, and a top that could be gripped to cut, dig, club, or hammer. It took skill and strength to fashion such an ax. Mostly they were made of quartzite, lava glass, chert, or flint that had to be broken out of larger stones and then honed to an edge, probably with smaller rocks, animal bones, or antlers. Since it was often roughly the size of a hand, it was transportable, and the fossil evidence indicates that it traveled with *Homo erectus* wherever he went.

And by all accounts, *H. erectus* went farther faster than any previous primate, probably because of his preference for meat. Our best guess is that he followed herds of herbivores that supplied him with food, clothing, and tools. While recent fossil finds indicate that *H. habilis* made some forays into the Middle East and southern Russia, *H. erectus* migrated out of Africa almost the moment he arrived on the scene, and immediately began to put thousands of miles between him and his African homelands.[10] Rutgers University geochronologist Carl Swisher III and his colleagues have found *H. erectus* sites in Indonesia and the Republic of Georgia that date to between 1.8 million and 1.7 million years ago. After that he headed far into China and eventually over the land bridges of Southeast Asia into Australia.[11]

But more than the tools, stature, and travel habits of *H. erectus* had evolved. In proportion to his body size, he had the largest brain of any primate, or any animal alive. It was a good two-thirds the size of ours, and some 50 percent larger than that of *Homo habilis,* crowding the forehead outward so the old apelike slope was nearly gone. His head and face would not have looked exactly like ours, but hints of us would have been there. The brow ridges were thicker, the mouth still a bit muzzlelike, but his would have been an expressive face. Given the environment he was living in, nature would have favored that adaptation because it made him a better communicator. Here was a creature, after all, who was not only smarter, but, arguably, more social and more interdependent than *H. habilis.* That would have made the communication of joy and sadness, aggression and anger, amusement and contentment, remorse and sexiness more important than ever as his inner world grew more complex and required better tools for socializing.

The ability to charm a mate, communicate during hunting, or stare down a competitor would have been invaluable because despite his prodigious intellect, *H. erectus* was almost certainly incapable of speech as we know it. Standing upright had helped rearrange the shape of his throat and larynx, but generally scientists agree that the relationships among tongue, lungs, throat, and nose still needed fine-tuning before speech was possible. All efforts to teach chimps or other primates to speak words have failed miserably, not because they aren't intelligent, but because they simply don't have the throats, muscle control, and nerves needed for it. (There is also a difference between spoken language and simply saying words. A parrot can say words but doesn't understand what it is saying or abide by any rules of grammar.)

Nor had Broca's area likely reached the level in *Homo erectus* it has in our species. Our brains are about 30 percent larger, and most of those additions have shown up in the cerebral and prefrontal cortex, areas key to high cognitive functions such as modern speech. Even if the cerebral horsepower *was* there to think far more sophisticated thoughts than any gorilla or chimp, and even if the muscles for speech were in place, the neurons that are needed to control the intricate musculature that speaking requires may simply not have evolved yet. Alan Walker has pointed out that the inner column of *H. erectus*'s vertebrae was probably too small to handle the rapid and concentrated neuronal commands that wordmaking requires.[12]

As evidence, Walker points to a small bone in the spine of Turkana Boy known as T7, the lowest vertebra of the thorax. All vertebrae have a hole in their centers through which the spinal cord passes. In modern humans this hole is large enough in T7 to fit all the nerve fibers needed for the fine muscle control of the rib cage and lungs when we exhale (something necessary to speech). But that same hole is smaller in Turkana Boy's spine, even taking into account his teenage size. Walker believes that the wiring for finely controlled speech simply wasn't capable of carrying all the signals needed to generate even rudimentary speech.[13]

Nevertheless, *H. erectus* was, as Walker has put it, "devastatingly clever for his time," and as his more advanced toolmaking indicates, he did have fast and facile hands. Perhaps he had other ways of communicating that didn't require words. If the F5 region of his brain was blossoming into a primordial version of Broca's area, and combining the power of his mirror neurons with a refined ability to manipulate objects, maybe *H. erectus* was able to create gestures, rather than sounds, that held meaning.

This would have been a shattering leap forward. Given the increasing complexity of his life on all fronts—environmental, emotional, mental, and social—the forces of natural selection would have favored any improvement over facial expressions as a way to share ideas and feelings, positive or negative.[14,15]

. . .

British psychologist Merlin Donald has speculated that the best way for *H. erectus* to communicate would have been to mime what he was thinking.

Fluttering fingers might have gotten across the idea of a bird or flying. Or the motion for building or using a particular tool might have become a perfectly natural way to communicate the idea of the action itself—a digging or a cutting motion, for example.

Some scientists have speculated that the first proto-word/gesture might have been a pointing index finger that said, simply, "there." Human toddlers spontaneously develop this ability at about the time they start to use their first words, around the age of fourteen months, the age, strangely enough, when a child attains a *Homo erectus*–size brain. In his book *The Hand,* neurologist Frank Wilson points out that cognitive and developmental psychologists agree that pointing in children is a "gesture of intentionality" that separates us from chimpanzees and is a unique milestone in human cognition. Chimpanzees do not spontaneously point with their index fingers, and they can't be trained to use them (at least not with any knowledge of what it means).[16,17]

H. erectus's miming wouldn't in some ways have been very different from the case of the habiline father "explaining" to his daughter how to fashion a flint knife simply by showing her the process. If chimp behavior is any indication, even by the time *Homo habilis* emerged our mirror neurons would have been plenty well equipped to make the connection between a mimed action and the idea of the actual thing. And this would have been even truer when the "devastatingly clever" *H. erectus* showed up on the scene.

Over time, with so many thoughts to share and threats to make and information to express, body language and facial expressions might simply have fallen short of communicating all of the increasingly detailed information *H. erectus*'s brain was generating. Gestures representing vital but easily expressed information might have been central to better representing a large predator or food or drink. And in time these gestures might have been adopted throughout the troop to mean the same thing to everyone.[18] Eventually, perhaps, random mimes began to evolve into more complex forms of communication—the first crack at a language, complete with a rudimentary syntax.

The problem with these theories is that they are difficult to prove. There are no fossilized artifacts that can shed light on how *H. erectus* learned to share his thoughts and feelings. Fortunately, however, some interesting theories have arisen out of the behavior of children.

Child psychologists (and most parents) know that even at eight months of age, a baby's thought processes are extremely sophisticated. They clearly have things they want to say and needs they want to express, but the shape of an infant's throat, and her immature brain anatomy and nervous systems prevent her from actually using words.[19,*]

Toddlers, however, *are* perfectly capable of gesturing. Waving bye-bye is the best-known universal example. New studies have recently revealed that with a little help, this natural ability to communicate with a gesture can grow far more sophisticated than originally thought.[20]

In the 1990s a researcher named Joseph Garcia noticed that healthy, hearing babies born to deaf parents who communicate using American Sign Language (ASL) began to talk much sooner than the children of hearing parents. The interesting thing was that they weren't speaking with their voices, they were speaking with their hands, just like their parents!

In subsequent studies Garcia also found that these infants would spontaneously sign that they were hungry, thirsty, or had a wet diaper as many as eight months before they were capable of actually saying a single word. This meant that the children's brains were developed enough to talk, but they had to do it with their hands because they couldn't do it using their throats.

At about the same time that Garcia was noticing this, Linda Acredolo at the University of California at Davis and Susan W. Goodwyn of California State University at Stanislaus also found that infants could communicate by hand gesture at eight months.[21] Acredolo concluded that some form of gesturing is second nature for most children. "It's just that people haven't paid attention, and parents are so focused on words that they don't see this as something to be encouraged."

Just as toddlers can imitate the gestures for "Itsy Bitsy Spider" or sniff a flower and then use that one sniffing gesture to indicate the experience of sniffing any flower, they also can learn to associate gestures with a need, a thing, or a concept. In one study a little girl started with simple signed words such as "milk" and "more," but within months was communicating far

---

* At about nine months a baby's throat elongates enough that her larynx drops sufficiently to begin to tackle speech. Several months later the cerebral wiring for transmitting the signals required to make words falls into place.

more complex ideas. During a visit to an aquarium when she was ten months old, she was watching penguins swim and signed "fish" to her mother. Her mother corrected her, and signed "bird." That confused the girl, who again made the sign for "fish." Then her mother signed "bird" plus "swimming." *That* her daughter understood. Just two months later, the same girl picked up a feather lying on the ground and signed "bird hair," an indication that she was now capable of combining two separate, previously learned concepts into an entirely new idea.

The work of Garcia, Acredolo, and Goodwyn all show that once exposed to signing, children take to it naturally because at this stage of their lives their hands are far more facile than their throats and mouths. As Elizabeth Bates, a pioneer in the field, put it, it is simpler to "imitate and reproduce something with a great big fat hand [than] the mini, delicate hundreds of muscles that control the tongue."[22] In other words, we don't speak at this early age not because we don't have the brainpower to think up ideas we want to express, but because we don't have the properly developed neural pathways, throats, lungs, and tongues needed to say words.

Acredolo's study even revealed that children who used gestures to communicate before they learned to speak developed IQs later in life as many as twelve points higher than children who didn't sign in infancy. That indicates that the earlier children can share what is on their minds, the better they later grasp deeper intellectual concepts.[23]

Whatever the long-term benefits, the beauty of learning to communicate by gesture at this young age seems to have been that both parents and children became happier. One mother reported that when she started to sign with her eleven-month-old son, the noise level in the house dropped dramatically—not because they were silently signing but because her son wasn't nearly as frustrated as he had been when his screaming and crying failed to make his point.

In fact, all three researchers found that toddlers often take rapidly to gesture because they are frustrated and can't tell their caregivers what they want. Grimacing faces, twisting torsos, and outright bawling simply aren't specific enough tools for a ten-month-old to say, "I'm hungry. I'm wet. This hurts. I want my bubby." Gesture, however, accomplished the job beautifully, once babies learned the basics.

If a toddler can gesture symbolically to share what's on her mind, could *Homo erectus* have done the same, especially given the survival pressures he was facing? He had the manual dexterity, the mirror neurons, and the social and practical needs. And he had a brain approximately the size of a toddler. This doesn't mean that *H. erectus* had attained a human toddler's intelligence—some aspects of brain architecture were certainly different—but it is possible that their general perceptions of the world were comparable. And both had strong reasons to communicate, even if they couldn't use words to do it.

These haven't been the only studies that reveal unexpected connections between gesture and language. A few years ago Laura Ann Pettito, a developmental psychologist at Dartmouth College, and her colleagues made a remarkable discovery when they were studying the babbling that all children do beginning at about the age of seven months.

Most scientists long ago concluded that the ba's and blah's and da's, the raspberries, and endless trilling and slurping toddlers do at this age mark the earliest stages of an infant's efforts to master the rhythmic, singsong sound patterns that lead to phonemes, words, and, eventually, sentences—the building blocks of language.

What made Pettito's experiments unique is that she, like Joseph Garcia, worked with babies whose parents didn't use words, but instead communicated in ASL. Her study focused on two groups of three babies each. The first group's parents spoke and heard normally. But the second group was made up of the children of deaf parents who used ASL. In both cases, all the children could hear normally.

As expected, each of the two groups began babbling at about seven months of age. But Pettito uncovered some peculiar behavior among the children of deaf parents: they were babbling with their hands as well as their mouths.[24] Pettito's team concluded this meant that the language patterns that the brain locks in during infancy are not exclusively related to sound. There are deeper rhythms the brain is grasping and reworking. Apparently if the child is exposed to spoken language, then he expresses those rhythms with sound, but if he is mostly exposed to sign language, then he expresses them with his hands.[25]

Hand babbling makes no more sense than the noises most talking babies make with their mouths; nevertheless, Pettito says, it represents a distinct effort to get a handle on the underlying rhythms of language, verbal or otherwise.[26]

Brain-scanning technologies have also recently revealed that Broca's and Wernicke's areas, the two parts of the human brain most directly involved in generating and understanding language, are as deeply involved in processing gestural language as they are in processing the spoken word.[27] Stroke victims often suffer forms of aphasia that can destroy their ability to speak sensibly or to understand speech spoken to them. Most victims of Broca's aphasia speak in gibberish, even though in their minds they think they are speaking with absolute fluency. If a victim is suffering from Wernicke's aphasia, someone may speak sensibly to them, but to the victim the words make no sense. In both cases the wiring that processes speech, incoming or outgoing, is tangled and destroys the brain's ability to parse language.

Precisely the same fate befalls ASL users who experience strokes that damage *their* speech centers, except in their case they lose their ability to sign, or comprehend signed language.[28] ASL victims of Broca's aphasia can sign, just as speaking victims can make wordlike sounds, but the signs, even though they resemble ASL, make no sense.[29] Apparently these areas of the brain don't see language in the form of sound alone, they process it as movement as well. Why? Arguably because those parts of the brain evolved the ability to understand gestures first, and spoken words later.

. . .

There is no more dramatic example of how hand gestures and mimicry might have once evolved into true language than the remarkable story of students attending two schools created to help deaf and mute children in Nicaragua following the 1979 Sandinista revolution. If scientists could somehow drop out of nowhere and watch the evolution of a completely new language from scratch, this is as close as they could hope to come.

During the eight-year civil war that waged between the Nicaraguan government and Sandinista insurgents, a new government was formed in 1985. Early on it initiated a program to help the nation's deaf, nonspeaking children. Two schools were opened in Managua, the country's capital, and children poured in from around the country to attend them. Because of the long war, none of these children had been trained in any of the world's two hundred known and accepted sign languages. At best they had only the rudimentary pantomimes each had developed out of necessity with their friends and families as they grew up—gestures for food or drink or sleep, but not much more.

Unfortunately, the teachers at the new schools weren't much help to their language-impaired students. They had been urged by Soviet advisers to teach the children to finger-spell, a system that uses a single gesture for each letter in an existing language's alphabet. The problem was, the children had absolutely no concept of an alphabet or words or any kind of language. They couldn't spell with their hands any more than they could spell with their mouths. But the children did have an overwhelming desire to communicate. And so they did something extraordinary: They began to talk among themselves, using their hands. First they started with the basic mimes they had brought from their homes. With these as a foundation they next began to create an entirely unique language, all while their frustrated instructors stood by and watched in amazement.

In June 1986 the Nicaraguan Ministry of Education asked Judy Kegl, an ASL expert, to come to the schools to help the instructors understand what was going on. Kegl figured she would try to get an overview of the ways the children were communicating and thought she might then compile a rudimentary dictionary of the symbols the children were using. First she visited older, teenage students who were taking a hairdressing workshop. Though Kegl could sign ASL fluently, it did her no good because the only signing these students used was the variety they had fabricated on their own. In fact, she quickly realized that the gestures they used, while creative, were nothing like ASL. In fact, they were pretty awkward and didn't seem to have anything like the underlying patterns you find in real speech, even gestured speech. It all seemed chaotic.

Many gestures—those for "eyebrow tweezers" or "rolling curler," for example—were simply represented by miming the act, the way, perhaps, *Homo erectus* might have mimed "big predator" or "fluttering birds." Some gestures were more complicated. One of the teenage girls showed Kegl a sign in which she laid her left palm flat, then with her right hand drew a squiggly line from the middle finger to the base of the palm. Then she turned her right hand over and pointed below the belt. Kegl didn't get it right away, but eventually figured out she had just seen the sign for tampon.[30]

What Kegl was witnessing was a kind of gestural pidgin or protolanguage based on the hand movements the children had brought with them from home. These were the nonverbal versions of Bickerton's Hawaiian-Japanese-English pidgin *"aena tu macha churen, samawl churen, haus mani pei,"* which explained why they had no real grammar or system of rules, at least not yet.

But just as Bickerton found that pidgins can transform themselves, sometimes within a single generation, into much more mature Creole languages complete with refined syntax and grammar, so did Kegl, except that she was shocked to find the salvation where she did.

After visiting the secondary school, Kegl made her way to San Judas, the primary school attended by the youngest deaf students. While there she noticed a girl named Mayela Rivas signing in the school's courtyard with a rhythm and rapidity she hadn't seen among the older children. She remembers thinking that the girl was using some sort of internal rule book as she signed. And in a sense she was.

The smaller children, it turned out, were pushing the old pidgin the older students had invented into new territory. Ann Senghas, who was a graduate student of Kegl's in 1986, later said, "It was a linguist's dream. It was like being present at the big bang." In a paper she wrote for the journal *Science*, Senghas explained that the younger, not the older, children were taking concepts, things, or actions and breaking them down into discrete units—gestural symbols—and creating a true language.

The older children, on the other hand, tended to use gestures that painted a kind of moving picture of an action. Rolling downhill, for example, might be shown with a gesture that indicates both the type or manner of the movement (rolling) *and* the direction or path (down) at the same time with a flopping hand or jiggling line moving downward—something you or I might do in conversation to illustrate a point. But for complex communication this was too complicated and difficult for others to accurately repeat. It was like saying a long, jaw-breaking word that has a very narrow meaning. Such a word won't very often be used because its meaning is so specific. It has no versatility.

The next generation of students didn't simply learn to perform hand movements better than the older children, they redefined them by breaking them down into smaller signals that could be easily combined with others to express more ideas. Rather than indicating "rolling" and "down" in one elongated sign, they instead created one sign, or word, for "rolling" and a separate gesture for "down." Rolling was indicated with a circular hand motion, quickly followed by "down," which was indicated by placing the hand at the chest and then cutting the air and straightening the arm with a clear, downward motion, almost like a salute.

This is much more the way both true signed and verbal languages work, unlike nonverbal forms of communication such as pictures or painting or mime. In speech, information comes in discrete bits like the sounds of letters and words that are then arranged in increasingly larger bits like phrases and sentences, all set in a certain order.

The other hallmark of language is that by breaking objects, actions, places, etc., into small pieces, communication becomes more flexible; each piece can be used and reused in different contexts to mean different things. This is how we transform a word such as "grip" from something that is a purely physical description, as in "I gripped the hammer," to a highly emotional one: "Get a grip!" This is the very essence of metaphor and simile and context, which make all languages so elegant and powerful.

In the case of the Nicaraguan children, once they redefined the gesture for roll, they were now free to combine it with any number of other gestures to say "roll up" or "roll over," or eventually, "I rolled the idea around in my head." This makes the gestures less iconic and more symbolic, less mime-like and more versatile. The two innovations—discrete parts and reusable parts—characterize the flexibility that makes a language of finite words capable of expressing infinite numbers of ideas, describe an infinite number of scenes, or tell infinite varieties of stories. Like DNA or the keys played on a piano, they can be combined and recombined again and again with strikingly elegant and dramatic results. Just consider the millions of species and billions of unique human beings that the four-word vocabulary of DNA has created, or the number of musical notes that have been recombined into pieces as diverse as "Row, Row, Row Your Boat" and Rachmaninoff's *Prelude in C Minor*.

By continually refining these hand signals over the past twenty years, each succeeding generation of young children at Managua's schools has improved their homespun language until now it no longer consists of crude, elementary gestures but a rich vocabulary capable of expressing every possible concept, from time and emotion to irony and humor. And what makes it truly amazing is the children managed this on their own without any central planning. No one sat down and wrote a grammar or created a dictionary. No one developed or taught a course. The language simply emerged naturally, out of the interaction of the children as they struggled to share what was on one another's minds.

As humans they were so moved and driven to communicate that they invented a full-blooded language, which today is officially known as Nicaraguan Sign Language, and they did it from the ground up without even really meaning to. Yet, governed by a couple of key but simple and unstated rules—take small pieces of information and then assemble them in a hierarchical, orderly, and sequential way—they invented a form of communication as expressive as any in the world.[31]

The plight of these children is as close as we will ever come to witnessing a language evolve from scratch. And it may provide a glimpse into the past. Perhaps like these children our ancestors had an overwhelming need to communicate, but neither the voice nor the language to manage it. Yet over time perhaps they found a way. After all, it seems that the underlying rhythms, syntax, and building blocks for expressing thoughts and feelings run deeper in the human brain than the structures for sounding out words and syllables alone. The mind is driven to share information. It begs for expression any way it can manage the work, as the hand-babbling of babies raised by signing parents and the remarkable story of the mute children in Managua reveal, and hands seem to be the tool of choice. For those among us who are voiceless for any reason, they quickly become manual vocal cords.[32] That could mean we were wired to communicate even before we could utter a single spoken word, and perhaps *Homo erectus,* or his descendants, mastered the form before he mastered speech, just as his forbears had learned to manipulate tools and weapons.

But powerful as gestures can be, and as much as they must have advanced and enriched the mental, emotional, and social worlds of our ancestors, they had their shortcomings. And before our predecessors could make their final leap to humanity, other adaptations awaited—among them one that was, so to speak, on the tips of their tongues.

III
Pharynx

Chapter 5

Making Thoughts Out of Thin Air

*Elmer Gantry . . . was born to be a senator. He never said anything
important and he always said it sonorously. He could make "Good
morning" seem profound as Kant, welcoming as a brass band, and
uplifting as a cathedral organ. It was a 'cello, his voice, and in the
enchantment of it you did not hear his slang, his boasting, his smut, and
the dreadful violence which (at this period) he performed on singulars
and plurals.*
—Sinclair Lewis, Elmer Gantry

*For it is a very remarkable thing that there are no men, not even the
insane, so dull and stupid that they cannot put words together in a
manner to convey their thoughts. On the contrary, there is no other
animal however perfect and fortunately situated it may be, that can do
the same.*
—René Descartes[1]

WE ALREADY KNOW THAT five million years ago the ancestral
apes from which we sprung found themselves left to survive on
Africa's expanding grasslands. And we know that led to upright walking.
But what we don't know is precisely why *Homo erectus* strode into the picture almost perversely optimized for running. Finding the answer to that

question, strangely enough, also illuminates how we became the talking animal.

For millions of years, evolution had been selecting for savanna apes of increasingly upright stature for several reasons, but the most compelling is probably that equatorial Africa was hot. A principle known as Bergman's rule states that in colder climates an animal will tend to be stocky and spherical to reduce the amount of skin it exposes to the air. The less body area exposed, the rule goes, the less heat it will dissipate, and so it will stay warmer. The inhabitants of Siberia, for example, tend to have shorter arms and fingers, even in relation to their shorter, stockier bodies to reduce the amount of skin they expose to the cold air.[2]

The reverse is also true. A taller and more cylindrical body will cool rapidly because more of its skin is exposed. Any elongated body will lose more heat this way, but ours is especially good at this because it is (1) naked and (2) perforated with sweat glands, some two and a half million, depending on your exact size. Thanks to these tiny glands, which cover almost every centimeter of our body, we can efficiently scatter 95 percent of the excess heat we generate.

This is an unusual approach to cooling in the animal world. Most mammals don't sweat much, they pant (possibly because they are mostly covered in fur, which restricts efficient sweating). And though other primates do have sweat glands—chimps, for example—humans have roughly twice as many.

Anthropologists generally agree that we sweat as profusely and efficiently as we do because if we didn't we would now be extinct. And the reason we would be extinct is that we evolved from scavengers to hunters. By the time *H. erectus* emerged we were well on our way to becoming the most cunning predators on Earth. We didn't have much in the way of claws and fangs, strength and speed, but we had the weapons we had created. And we could run.

No other creature is a better long-distance runner than a human being. Cheetahs may accelerate faster, ostriches and horses may gallop longer at high speeds, but no animal can cover more ground, without rest, than we can. The popularity of distance, marathon, and ultra-marathon running is a legacy of that ability, but there are plenty of other exotic examples of our peculiar bipedal talents. The Tarahumara Indians of northern Mexico, for example, routinely hunt deer by literally running them down over a period of days. They rarely get a close look at their quarry, but manage to stay near enough that the animal never has a chance to rest. Eventually the deer collapses from exhaustion, sometimes with its hoofs worn to nothing. The hunters, on the other hand, though tired, are far from dead.

The Nganasan people of Siberia use a similar method to hunt reindeer. They may follow an animal for six miles or more, hiding behind trees and piles of rock until they get within striking distance. Then with a burst of speed they close in and kill the deer. The Aché of Paraguay, the Agta of the Philippines, and !Kung San of the Kalahari Desert also use their legs and lungs to wear out their prey.[3]

This isn't the only way to bring down supper, but *H. erectus,* so perfectly engineered for running as he was, almost certainly used the technique as one of several weapons in his arsenal. However, in a world of intricately evolving feedback loops, this would have created yet another problem. The solution to that problem may help explain why the brains of our ancestors expanded so suddenly between a million and two million years ago, laying the foundations for a mind capable of speech.

. . .

Brains are greedy and costly organs. Compared with the rest of the body, they gobble up energy and burn hot. The modern human brain consumes as much as 25 percent of an adult's daily energy needs (more than twice that of chimps or gorillas). Ounce for ounce it devours sixteen times the calories that muscle tissue does.

Homo erectus's brain was not as large as ours, but it was getting there, and by some estimates was burning up to 17 percent of its body's energy budget.[4] On the hot savanna an animal, especially one with a brain that large, running long distances would have grown very warm—so warm that it almost certainly would have collapsed from heat stroke, no matter how tall and no matter the number of sweat glands it had, unless it found other ways to stay cool.

Many of our ancestors probably did die eking out a living under Africa's blistering sun, but some, our direct ancestors in particular, obviously didn't. Why? Because according to one fascinating theory they developed a genetic mutation (or perhaps several) that provided them with an ingenious air-conditioning system, one we still enjoy today.

Dean Falk, an anthropologist at Florida State University, believes that when our precursors first began scavenging for meat about two million years ago, they also began to evolve a network of cranial veins that cooled the blood running through their brains, faces, and skulls. She calls this the "radiator hypothesis" because the network operates something like a car radiator.

When we begin to overheat, the heart pumps cooler blood from the body and face into a fine network of "emissary" veins scattered in tiny branches throughout the skull near the scalp. Here, more heat escapes before the veins return air-cooled blood back to the brain, where it replaces the warmer blood that is already there.[5,6,7] In other words, it's a perfect natural radiator.

Falk developed this theory after closely comparing the skulls of modern apes, humans, and our ancestors. Though the veins and arteries that were once a part of these skulls were long gone, the skulls still revealed some of the vascular passageways that these creatures used. She found that we and apes have very different ways of moving blood to the brain, especially when it begins to overheat. Apes do not have nearly the complex system of heat-radiating emissary veins we do, and they have less effective ways, generally, of pumping cooler blood into the brain.

This also seemed to be true of early australopithecines such as Lucy, and evolutionary dead-end species such as *Australopithecus robustus,* both of which had more in common with jungle apes than with us. But the skulls of more recent species such as *Homo habilis, Homo erectus,* Neanderthals, and early *Homo sapiens* suggested that the emissary vein system progressively increased in size and complexity as the brains grew larger and as cooling requirements became more demanding.

In our species, Falk found the system rich and efficient, capable of radiating the majority of our body heat when we have worked up a sweat. (When you see the steam billowing off of a football player's head on a frigid Sunday afternoon after he has taken his helmet off, you are witnessing this system at work.)

Falk's view is that this cranial air conditioner coevolved with all of the other cooling mechanisms we were developing: the loss of our fur, our increasingly upright posture, our proliferating sweat glands. But it was especially crucial because without this adaptation the size of our brains would have been stunted, unable to increase much beyond *Homo habilis* size for the simple reason that any hominid on the savanna that didn't enjoy its benefits would have died of heat stroke long before it had an opportunity to pass its genes along.

And perhaps this is why some lines of savanna apes *did* die out: they never developed a truly effective cooling system that enabled them to take up scavenging and, eventually, hunting under the savanna's broiling sun. Hyperthermia can be a very efficient killer. A rise in the temperature of the human body of only four or five degrees Celsius above normal can scramble brain functions in humans that cause delirium, hallucinations, and convulsions. Vascular physiologist Mary Ann Baker has even written that the temperature of the brain may be "the single most important factor limiting the survival of man and other animals in hot environments."[8]

So while the same old evolutionary pressures would have been dogging our ancestors, favoring greater tool use, better communication, and all of the increasingly complex behaviors that define our species, without the evolution of some sort of brain cooling system, these advances may well have found themselves cerebrally stuck in neutral, doomed at best to have left our ancestors as unchanged as their chimp and gorilla cousins.

On the other hand, if the system did evolve and work, as it seems to work in us today, then the lid, so to speak, would have been off. Now the brains of our ancestors would have been free once again to expand. With their evolving ventilation system, *H. erectus* would have been able to hunt the herds he followed, *and* develop the intellect needed to face the increasingly complex social interactions of the troop, until eventually his hominid brain evolved into the unusually large size we see in us today. (The human brain is about three times the size scientists would expect to see in a nonhuman primate of equivalent body size.[9])

This made our ancestors' blood-cooled brains, in Falk's words, "a prime releaser," if not a prime evolutionary mover. By that she means that this adaptation doesn't stand with thumbs and toolmaking as a sea change in human evolution, but it enabled our ancestors to supply their reevolving brain the ability to grow still larger.

Exactly how would this new capacity be applied? If *H. erectus* was capable of some sort of gestured communication, but not yet capable of true speech, maybe it enabled certain parts of the brain already largely devoted to manual dexterity to grow and commandeer the hundreds of muscles and organs originally evolved for breathing and eating so that they could also meet the pressing need for increasingly refined and subtle communication.

It may have enabled the brain to fully build out the raw neuronal power needed to transform our ancestors from the very bright, but largely mute, apes they were into the smooth-talking species we are today.[10]

But even as that happened, certain other reorganizations also had to take place in the elongating throats of the tall and lanky creatures that were now rapidly spreading across the planet. A new, strangely shaped chamber in their necks needed to develop, an organ called the pharynx. Because without it, speech would be impossible.

. . .

The pharynx is cone-shaped and about four-and-a-half-inches long. It sits right behind the root of our tongues and connects our mouth to our esophagus. Strangely enough, the human pharynx evolved, at least partly, because we took to running upright. When our ancestor hominids stood on their hind legs, their necks slowly began to straighten and elongate. Over time their shoulders and torsos centered under their heads, their brows grew less sloped, their jaws more square, and their skulls more rounded. All of these changes caused the roofs of their mouths to rise; their necks to stretch; and, most important of all, their tongues and larynx, or voice box, to drop farther down their throats.

Others animals have a pharynx, but the architecture of ours and the organs positioned within and around it make it unique; a specialization as odd as the necks of giraffes or the strange binocular eyes of hammerhead sharks. Despite some of the work done with the likes of Koko the gorilla or Kanzi the bonobo chimpanzee, no other primate has been blessed with a pharynx capable of the noisemaking legerdemain ours possesses.[11]

The descent of our larynx created a ticklish problem for our ancestors. The throats of other primates are arranged so that their nasal passages are connected directly to their lungs by a single airway, while another, separate route links their mouths directly to the stomach. These run like two parallel, nonintersecting roads from the skull, down the neck, and into the torso, and never share a centimeter of common real estate.

This was probably the case with all of our ancestors as well, until, perhaps, H. erectus. With H. erectus, whose upright alignment was virtually identical to ours, the shape and length of our skulls and necks likely forced our nasal passages and mouths to forsake their formerly separate routes and join one another, creating an intersection in the back of our throats. And therein

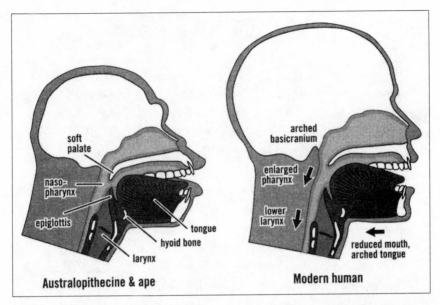

These two diagrams show the throats of apes and australopithecines (left) and modern *Homo sapiens* (right). Unlike other primates the airway and esophagus of humans intersect. As a result we can choke, but we can also speak. (Reprinted from *The Symbolic Species* by Terrence Deacon, used by permission of W. W. Norton.)

lies the problem, because the formation of that intersection meant that the food and water coming from our mouths could cross paths with the air we were breathing. And choking was born.

Increasing the chances of choking wouldn't seem to be an optimal evolutionary event. Even Darwin was surprised at this mutation. In fact, he seemed almost annoyed when he wrote in *Origin of Species:* "the strangest fact [is] that every particle of food and drink which we swallow has to pass over the orifice of the trachea, with some risk of falling into the lungs."

To handle that danger, we have a small flap of skin and cartilage called an epiglottis, which folds over the top of our trachea to prevent food and liquid from free-falling into our lungs when we swallow. This small organ does occasionally fails us, however. Until the invention of the Heimlich maneuver, six thousand people a year died from choking in the United States, usually while talking and eating at the same time. This made it the sixth-leading cause of accidental death. A chimpanzee, however, will never choke to death, at least not because the banana it was eating went down the wrong pipe.

An exquisite series of events unfolds every time our lungs pass air up through our throats where we bend, bite, and twirl our breath to rattle off a word as simple as "bread" or as convoluted as "supercalifragilisticexpialidocious." Commanding all of the apparatus needed to speak with the easy, unconscious fluency each of us manages is a remarkable feat. The body and brain recruit over one hundred muscles to do the job, more than any other human, mechanical activity. When we speak we transmit twenty to thirty phonetic segments or six to nine syllables per second. This takes tremendously refined breath control, specialized muscles that can expand and contract with unparalleled rapidity, and fibers in the tongue that enable us to move and reshape the air we exhale with lightning speed.[12]

Not that scientists have a complete grasp of how these elusive processes work. They know that the human cerebral cortex— the largest part of our brain—has more direct control over the face, tongue, larynx, and lungs than any other mammal or primate, and a lot of that neuronal firepower is used when we speak, but they don't understand the processes in all their detail.

In most mammals, facial expression, breathing, and the muscles of the mouth and throat are controlled by clusters of neurons called the reticular premotor area, an ancient part of the brain stem that is directly connected to our spinal column. In all animals the reticular premotor area controls many of the body's unconscious, visceral activities, such as swallowing, blinking, or breathing.

But as primates evolved, more and more neurons called "pyramidal cells" (so called because of their triangular shape) sent out long axons from the rapidly evolving cerebral cortex to connect directly and deeply with the nerve systems that manage the lungs, larynx, face, and tongue.[13] This eventually gave us the more conscious control we needed to trip the switches in all of those organs we use to form the words that express our thoughts.[14]

When you bend over to pick up a big box, you instinctively stiffen your torso as a brace against the weight by trapping the air in your lungs beneath a couple of retractable but very strong muscles called the vocal folds (known to most of us as the vocal cords). These cap your voice box or larynx, the front end of which you see in the mirror every day in the form of your Adam's apple (women less than men). Exert yourself a little too much and you will grunt because the effort pushes a little of the air out of your lungs and through the folds.

We use the same system to speak, except with considerably more subtlety. Speech begins with inhaling the amount of air our brain has calculated we will need to say what we want to say. Once in our lungs, we then release it in controlled puffs up through the windpipe (trachea) until it collides with the vocal folds at the top of larynx. It is here that we first begin to shape the air to make specific sounds. We may, for example, want to make a buzzing noise. If we do, we produce a zzzzzzz; if we don't, then we produce an ssssss-like sound. When you sing or hum or talk, it is the air rattling past these folds in rapid bursts that becomes the foundation for what everyone recognizes as your particular voice, a sound so distinct that it cannot be truly duplicated by any other human.

The tighter the folds of muscle, or the smaller and more rigid they are, the higher the timbre of your voice. If the folds are looser or larger, your voice will be deeper. This is one reason why big men generally have lower voices than little girls, although the shapes of our throats, nasal passages, and mouths also have a lot to say about whether we speak in rich baritones, smoky whispers, or nasal twangs. If we get excited and constrict our vocal cords, our voices rise. You might notice that when you are nervous, your voice sounds a little higher than it might normally sound.

While the larynx gives our voices pitch and character, we sculpt sounds into phonemes after they have made their way through our vocal cords and up into our throats. The English language consists of 40 or so phonemes.[15] They can be connected in innumerable ways to form every word in the language and plenty that don't yet exist. (Remember how many new words Shakespeare brought to the English language, all of them pronounceable.) Other languages also use set numbers of phonemes. German has 37; Japanese, 21; and Rotokas, the language of East Papua, a mere 11. No language uses more than 141 phonemes because that represents the outer limit of the sounds we can utter.[16]

Whatever language we speak, our tongues, lips, and teeth fold, push, and cut the sound waves to shape them into word-making phonemes. First the vibrating air reaches into the rounded chamber of our pharynx. Then we use the root, hump, and tip of our tongues to compress sounds or release them, or simply impede them before they escape from our mouths. We form the sounds for "gah" or "kay" or "tee" this way. Say almost any word (read these sentences out loud) and you can feel the rapid movements in your throat, tongue, and lips, all acting in perfect, linear synchrony.

Our lips are the last part of us to form word sounds. And we use them almost as deftly as we do our tongues. Sound out for yourself the subtle differences among "ef" and "vee" or "pee" and "bee." Only the faintest shifts in our lips make the differences among these sounds. We have this capability because our ancestors developed unusually fat and sensitive lips, well adapted for trying out jungle fruits and their flavors. It is pure evolutionary serendipity that they also turn out to be very good for making extremely explicit sounds.

But the connection between flavor and communication might be more literal than we once suspected. Scientists who have studied mirror neurons have found that when one monkey watches another eat and smack his lips, he, too, will tend to shape his lips into a smacking shape, instinctively reflecting what he is watching, as if somehow he, too, is eating. We do a variation on this when we find ourselves pursing our lips as two lovers lean forward to kiss in a movie. We do it as well when we instinctively try to finish a sentence for someone who is struggling to find the right word.

All of these realignments were, of course, accidental in the ways that evolutionary mutations are. However, if we had hadn't developed our peculiar pharynx, and traffic cop epiglottis, we couldn't hope to utter a single word.

This means our pharynx enabled our leap from gesture to speech, but the leap wasn't as far from gesture as it might first appear. An increasing number of scientists are concluding that somehow the areas of the brain that once exclusively coordinated the manipulation of our hands in space, expanded and evolved to also take on the duties of encoding thoughts into sounds and then manipulating the more than hundred specialized muscles needed to hold a conversation. In other words, our predecessors' brains learned to apply the same fine control over muscles in their throats that they had already developed to control the intricate muscles in their hands. Does that mean that in time our ancestors learned to gesture, not simply with their hands, but also with sound, using the muscles of their chests, throats, and mouths to manipulate air and make words?

There is no universal agreement on this question. Though every day we grow more adept at mapping the brain, those maps have yet to clearly reveal how language works. Part of the reason is that so many brain cells are crammed within such an exceedingly small space. Cognitive scientist Steven Pinker has pointed out that this could mean various parts of the brain dedicated to handling specific aspects of speech exist as small areas scattered all around the brain. "They might be irregularly shaped squiggles, like gerrymandered political districts. In different people, the regions might be pulled and stretched onto different bulges and folds of the brain."

PET scanning and functional magnetic resonance imaging (fMRI) do find that as we process speech, areas throughout the brain "light up." Of course, precisely what causes them to light up is far from clear, partly because language is itself so complex and interconnected that it's difficult to isolate a single aspect of speech, and partly because the brain is so convoluted, twisted, and turned.

Speech requires brain cells to coordinate the movement of the lungs, throat, mouth, lips, and facial muscles. It demands that we listen, symbolize concepts from both the outside world and the chambers of our own minds, and then apply phonemes in rapid succession to place words in just the correct sequence to deliver a sentence. Saying simply "How are you?" takes immense skill and brainpower.

The Music of Language

Our vocal folds also control the intonation of our voices, the modulations, intensity, expression, and volume that give them personality and add subtle meaning to the words we utter. Linguists call this prosody, a kind of body language for speech. We express it in the growling, the snarling, the singing, or the cooing we fold around our words. In conversation we absorb it in the tones, the rhythm, and the speed of the voices we are listening to.[17]

Great orators and communicators use prosody to pass along the subtleties of what is on their minds with the tiniest inflection, the smallest modulation, or the sudden hammering of a word so that rings in our ears and hearts like a bell. We all have this ability and we all use it to add new layers of power and meaning to what we say. It is another level of information that brushes our words with sympathy, doubt, confidence, anger, or sorrow. In this sense it is more closely associated with primal calls and cries, imbuing our voices with a kind of music. And the music we make is key in gaining and keeping the attention of others when we are speaking. In fact, the life we give to the words we speak, the emotions, color, weight, and speed we attach to them are central to the thing we call our personality, the cumulative impression we leave on others that distinguishes us from everyone else. It is central to who we are.

The emotional and musical side of prosody may explain why, unlike the words and syntax of language, it is processed in the right hemisphere of the brain rather than the so-called verbal left side.[18] The left side specializes in the sequences of actions, but the right handles shape and space. So in some sense the right hemisphere of the brain must "see" the intonations of the speech in terms of form and distance: imaginary objects that are close or far away, big or small, tall or flat, colorful or not. But even though different parts of the brain handle different aspects of speech, the content of words and the sound of words are not really separate. Together they combine to create, the meaning that inextricably binds ideas to emotion.

This ability is a testament to the extremely interconnected nature of the brain. With the evolution of the neocortex, the brain needed to send out longer and longer neuronal tendrils to keep new sectors in touch with more ancient ones.

The basal ganglia, for example, comprise an ancient section of our brain, an area that in other animals is exclusively devoted to movement. In reptiles, other mammals, and humans, it influences postures that display dominance and submission or attract the attention of the opposite sex. In us it still handles movements as basic as the way we swing our arms when we walk. We don't normally swing our arms forward and backward in unison because as far as our basal ganglia are concerned, we are still walking on all fours. It is a legacy of the way our ancestors moved before we stood up.

If a man puffs out his chest or a woman flutters her eyelids or you find yourself crossing your arms in a meeting when you hear something you don't agree with, your basal ganglia are hard at work. These nerve fibers also affect the expressions of your face—surprise, anger, confusion, interest. In a singles bar basal ganglia must be near overheating from input and activity. If there's not enough dopamine operating in your basal ganglia, you may shuffle and not swing your arms at all, an incipient warning of a disease such as Parkinson's. On the other hand, too much dopamine has been linked to the physical tics in Tourette's syndrome or obsessive-compulsive disorder.

The basal ganglia and other ancient and visceral parts of our brain are also wired into Broca's area. Unlike Wernicke's area, Broca's doesn't import information. Instead, it controls the expression of what is on our minds, even when we don't always consciously realize that those things *are* on our minds. It surrounds and serves and affects parts of the brain that handle body language and the gestures we make when we tell a joke or the expressions on our faces when we are talking with our significant other.

But in human beings, perhaps the most important operation the basal ganglia perform is to help us do the physical work of shaping thoughts into words and sounds before we ever part with them. In Broca's area we literally talk to ourselves, rapidly preprocessing all of the language before our basal ganglia pass the signals onto our throats, tongues, lips, and lungs to manufacture the thought-sounds we send into the world for others to hear.

It would be nice to know when and precisely how all of these capabilities fused, but we don't. We have no *Homo erectus* brains to examine or scan, no early *Homo sapiens* we can test to learn if they spoke in full sentences or the grunts B movies have imagined they did. But we do know from the mirror neuron work that Rizzolatti and another of his colleagues, Michael Arbib, a biologist and computer scientist at the University of Southern California, have done that other primates possess areas of the brain—the F_5 region, for example—that are located in roughly the same sector as Broca's area in the human brain. In us these areas handle both speech and manual manipulation, which, unlike the uncontrolled cries and calls of wild animals, are things we do intentionally and very consciously.[19]

In humans these two areas sit, cerebrally speaking, cheek by jowl. In fact, in 1998, Rizzolatti and Arbib wrote, "This new use of vocalization [in speech] necessitated its skillful control, a requirement that could not be fulfilled by the ancient emotional vocalization centers. This new situation was most likely the 'cause' of the emergence of human Broca's area."

More recently Arbib has concluded that *H. habilis* and even more so *H. erectus* enjoyed the benefits of a "proto–Broca's area" based on an F_5-like precursor. This part of *H. erectus*'s brain might have handled communication that was partly manual and partly facial and oral. In time, Arbib suspects, this early version of Broca's area gained primitive control of the vocal machinery as far as it had evolved. It began to play a puppeteer's role, except in this case it made the pharynx dance, not the hands and fingers. And once that happened, language and the brain would have nudged one another's evolution along, back and forth, as our ancestors experimented with gestures of sound rather than body, creating increasingly complex social situations that accelerated their own need to improve communication.

A crucial part of that progression would have required the development of increasingly refined control of the mechanics we use today to speak—not an inconsequential accomplishment.[20] It doesn't matter that gorillas and chimps have the mental capacity to symbolize some basic concepts (a little more than a hundred by last count). The problem for them is that no amount of coaxing or training allows them to do it with words, for the simple reason that they do not have the basic anatomical tools for it.

If we were somehow to discover William Shakespeare himself in the form of a great silverback gorilla wandering the misty forests of Rwanda, he would be entirely incapable of uttering a single line of *Hamlet.* But if Shakespeare had existed as *Homo erectus,* he might at least have managed a partial sentence complemented by several eloquent gestures.

How *Homo erectus* and the early *Homo sapiens* who followed might have refined their ability to speak as they slowly gained greater control over their lungs and voice boxes, lips, and tongues might have been similar to an evolutionary form of babbling. At birth, infants' throats are virtually identical to our ape ancestors'. They have one direct passage to the lungs and another to the stomach. During the first three months of life this enables them to breathe and nurse without any risk of choking. But at about three months of age, the larynx descends in their throats and creates our uniquely human pharynx. That provides room enough for the tongue to move forward and backward and to form all the vowel sounds we can make as adults.

During the next several months infants begin to compare the sounds they can make with the language they hear around them. A feedback loop rolls into place, and out of this give-and-take, a vocabulary begins to emerge that is expressed within the framework of the grammar and syntax that seem to be hardwired into the human brain.[21]

This transformation takes place without anyone sitting down with the baby and pulling out a rule book that explains what words are or how grammar works. It is a natural process and it happens universally.

At eighteen months something especially remarkable takes place. All of the foundations for speech seem somehow to simultaneously and miraculously lock into place. After months of physical and cerebral development, the mechanical and neuronal engines are assembled, wired, and ready to roll. We understand our first words just before our first birthday and begin to *say* our first words shortly afterward, but it takes another six months or so before the next phase of the process gets under a full head of steam. The basics of grammar slip into place. We can say sentences that exhibit simple syntax, such as "Me want that." And with that underlying template in hand, we begin to build our personal repertoire of symbolic noises—the things we call words—with astonishing rapidity.

During the ten years that pass between eighteen months of age and adolescence, children acquire an average of eleven new words a day, about forty thousand total, or one every two hours, a phenomenon that is repeated nowhere else in nature. This whole process is so powerful that it is as impossible to keep children from learning to speak as it is to prevent them from learning to walk. You would have to go to extraordinarily cruel measures to rob any human of the gift of speech.

. . .

How we managed as a species to get past simply making noises to the true creation of language is one of the great mysteries of science. Arbib looks at it this way: Something as complicated as human language would have required a precursor to hang its hat on. That brings us back to manual gesture. The experiences of the deaf children in Managua provide some real-world support for this theory. The first, older generation of students brought with them the iconic gestures they had struggled with at home, pantomimes that helped them communicate basic information. But it was the younger children who conceived ways to break gestures down into bite-size, reusable symbols. That was key.

Arbib believes that ancestral hominids like *Homo erectus* may have used similar pantomimes that later provided a core, if inefficient, gestural vocabulary that could then be modified and broken down into a growing stock of signed words. In this way he imagines that our ancestors' early iconic gestures would have evolved into the more efficient and complex signing that eventually enabled early humans to assign a symbolic sound gesture to an already existing symbolic hand gesture.

Or, according to Arbib, perhaps something slightly different happened. Maybe the first hominids, struggling to move from gestural language to speech, repeated vocally what the older children of Managua did with gestures: create a single word to represent a series of fairly complex actions. The sound " 'grooflook' or 'koomzash' might have once encoded a complex description like 'The alpha male has killed a big meat animal with long teeth and now the tribe has a chance to feast together. Yum, yum!' or commands along the lines of 'Take your spear and go around the other side of that animal and we will have a better chance together of being able to kill it.' "

If we combine Arbib's theories with the experience of the children in Nicaragua some interesting possibilities arise. Though that one word may say a lot, it is useful only in that single situation. It has no flexibility. So as words go, it doesn't do its job any better than the early Nicaraguan sign for rolling downhill. After a while our ancestors would have found themselves buried in stand-alone superwords with limited use. To escape this fix they may then have broken down the superwords into conceptual chunks that could be reused and reassembled into different meanings depending on their context; precisely the way the younger Nicaraguan children did with their gestures. Arbib imagines that a tribe of protohumans might have agreed on a sound that stood for fire. Later members of the tribe might have come up with additional sounds they agreed meant "burn" and "cook" and "meat." Soon simple, efficient sentences could be communicated that had very different meanings depending on how they mixed up the vocabulary. "Fire cook meat," or "Fire burn!" or "Burn (the) cook!"

Or take another example: A troop comes up with separate words for "ripe apple," "ripe plum," and "ripe banana," each word standing on its own. But later, if separate words evolved for "ripe" *and* each of the fruits, rather than three fruit-adjective words with limited uses, our predecessors would have found four words they could have used to say the same thing: "ripe," "banana," "plum," "apple," and ripe anything else, including "ripe old man," or a person "ripe" for the picking.

Was the development of true language built on this scaffolding? We can only speculate, intelligently. Perhaps with the arrival of our species 195,000 years ago, the transition from gesture to the spoken word was made, and the blossoming of human speech and culture began to gather speed.[22,23] Perhaps.

But for that to happen, something else deeply bound to the emergence of language would have had to have surfaced first: an awareness of ourselves—the understanding that we existed.

Chapter 6

I Am Me: The Rise of Consciousness

*The first human wave was, however, a little wave, threatening to
vanish. . . . Tremendous bodily adjustments were in process, and, in the
low skull vault, a dream animal was in the process of development, a user
of invisible symbols. In its beginnings, and ever more desperately, such a
being walks the knife-edge of extinction.*
—Loren Eisley, "The Angry Winter"

THE PREFRONTAL CORTEX is the newest part of our brain, the part
that sits right behind our foreheads. In evolutionary terms it developed
at a remarkably rapid pace. Three hundred thousand years ago, when the
last *Homo erectus* met his fate, it basically did not exist. Today every human
has one, which means that the most complex part of our brains evolved and
increased our overall brain size 25 to 30 percent in an evolutionary blink.

The prefrontal cortex is where we do most of our high-end thinking. It
is where we worry, symbolize, and process a sense of self and of time,
where we recall complex memories and imagine events in the future that
haven't yet happened.[1]

Cognitive scientist Terrence Deacon has pointed out that one of the
many reasons why the prefrontal cortex is so central to the human experience is that it is deeply wired into every other area of the brain, even extremely ancient ones. This makes it a kind of general contractor, keeping

the big picture in mind while staying in close touch with nitty-gritty work such as hearing, moving our limbs, and controlling our breathing. While other parts of the brain might tend to specialize in their fields, the prefrontal cortex is a generalist that plays a role in nearly all cerebral experience.

This area also houses an ability that is far less developed in other mammals and primates, something scientists call "working memory." We know that humans can symbolize thoughts and ideas, but working memory enables us to take a thought or a memory, set it aside, shift our attention to something else, then pick up where we left off. As I write this, I am thinking of examples I can provide that will explain what working memory is so I can get them down on paper. If the phone rings, I can set those ideas aside, answer the phone, have a conversation, and then haul the examples I was developing back out of my memory and further develop them.

This might seem simple. After all, we all do it all the time. But it is not simple. Not only are we symbolizing, encoding, and cerebrally packaging those thoughts or experiences, but we are also then recalling them, reconnecting them with all of the thoughts with which they were previously linked, and then conceiving new ways to develop them. It is as though they were objects we have made and shaped that we literally set aside while we take another object in hand to shape, and then we connect them.

Several remarkable abilities flow from working memory. First, it helps us prioritize and take better charge of our lives. If we can set knowledge aside and then later recall and reuse it, it follows that we can also decide to deny ourselves something in the short term to accomplish something more important to us in the long term. We might want a second piece of pie after dinner, for example, but we might also understand that having that pie will thicken our waistline and raise our cholesterol. So for the sake of our health, pleasurable as pie would be in the short term, we forgo it. Or we might go into debt to earn a master's degree, planning that later we will land a better-paying job that enables us to repay the debt and live a more fulfilling life in the bargain. Brain scans have shown that when we decide to delay one action in favor of inaction (in other words, when our working memory prioritizes concepts), sections of the prefrontal cortex activate.[2]

The prefrontal cortex's talent for prioritizing and inhibiting is one reason why we don't walk into parties and begin sniffing one another the way dogs do at the park. The prefrontal cortex is subduing the part of the brain that wants to act like a dog and guiding it to do more socially acceptable things, like smiling and shaking hands.

In addition to brain scans, we also know that the forebrain plays this inhibitory role because scientists have studied people who have had this part of their neocortex injured. Neuroscientist Antonio Damasio and his colleagues at the University of Iowa, for example, have described the cases of two people whose forebrains were damaged in infancy: a man by a tumor discovered at three months of age, and a woman who was run over by a car when she was fifteen months old. Both survived their injuries and went on to grow up in stable homes, with educated parents and healthy siblings. Nevertheless, both eventually began to have problems. As teenagers they stole and lied and generally seemed to have lost their moral compass. Though they were often pleasant, they would do and say awful things and then show no remorse for their actions.[3] Their forebrain seemed incapable of inhibiting some behaviors.

The most celebrated case of forebrain damage is the story of Phineas Gage, a railroad foreman who was tamping an explosive charge with a metal rod in Cavendish, Vermont, in September 1848, when the charge accidentally exploded, propelling the inch-and-a-half-wide, thirteen-pound tamping iron through his left cheekbone, behind his eye, and completely through his skull, destroying the front, left side of his brain. Amazingly, Gage survived the accident and even remained conscious as his coworkers got him to his feet and then to a local doctor named John Harlow. Harlow treated him so successfully that Gage went home to New Hampshire ten weeks later.[4]

Within a year, Gage actually felt well enough to work, but he was unable to land a job with his old employers, not because he was physically handicapped, but because his personality had changed. Before the accident, Gage had been one of the railroad's best construction foreman: capable, a clever problem solver, pleasant with his workers, and blessed with a shrewd sense of business. But now he was surly, grossly profane, and showed little respect for anyone he worked with. In fact, his fellow workers said he was "no longer Gage," but some other man who was stubborn, capricious, and impatient.

Lobotomies that sever prefrontal connections with the rest of the brain have sometimes had similar effects. They were performed by the thousands in the 1940s to treat everything from schizophrenia to criminal behavior.

They often disposed of personalities as well as patients' disabilities, but many times, rather than calming down these unfortunate people, it made them agitated and disruptive, like Phineas Gage.

Drinking is a common disinhibitor that works against the prioritizing capabilities of the forebrain. Tests show that alcohol affects the levels of the neurotransmitters dopamine and gamma-aminobutyric acid, or GABA in the prefrontal cortex. After the first drink or two, dopamine levels rise and increase the feeling of elation and confidence. It helps transform quiet people into gregarious ones and can make the shy vivacious. GABA slows neurons from passing along signals that either excite or inhibit other neurons. Since it slows down these activities, it also means that your prefrontal cortex is less likely to stop behavior like putting on a lampshade or dancing on the bar, which is why people sometimes do these things or get into brawls because someone looked at them the wrong way after too much to drink. Long-term research has found that chronic alcoholism affects prefrontal cortical capabilities like problem-solving and prioritizing. The point is that without working memory and the ability of the prefrontal cortex to delay, inhibit, prioritize, and basically generate its own mental cues, we would not act in a way most of us call human.

As it turns out, these same capabilities are central to our ability to symbolize and organize language. That, at least, is Terrence Deacon's argument. The symbols we attach to our thoughts make them portable and modular so that they can be looked over and rearranged like LEGOs. But the nature of holding multiple symbols in mind at once means that they also must be compared and prioritized for the simple reason that every one of them can't simultaneously be top of mind. Some have to be rapidly shuffled to the side and subordinated to others.

All of this makes yet something else that is extremely rare in other species possible for us. While the prefrontal cortex is busy processing sights, sounds, and smells that rattle inward along the myelin-encased highways it uses to reach into the deepest recesses of the brain, it is also manufacturing its *own* input, creating *new* symbols and new relationships *between* those symbols, without any stimulus from outside of our own mind. The brain is self-generating its own symbols.

These form still more new patterns that we "see" before making a decision about what to do next. This is what most of us call thinking—the conscious, deliberate kind. It makes us very unlike other creatures, who, like my dog

Jack, react instinctually and serially to whatever happens around them. In Jack's case he may be sleeping one minute, then the next, when he sees me put on a jacket, runs to the door to bolt outside with his nose to the wind, waiting for the next thing to react to. Jack does not juggle multiple experiences, comparing one to the other to decide whether he should run after the Frisbee, sniff the elm on the left, or urinate on the oak to the right. Jack's mind is incapable of purposefully prioritizing, or as John Locke put it, "Beasts abstract not." He just follows his nose and disposes of his experiences as they strike him. Deacon calls this indexical memory; a linear index of experiences or reactions that come in one lobe of the brain and out the other.

The Brain Explained

Of course, the brain can't be explained—that's what makes it so mysterious. It is a Rube Goldberg machine, a mishmash of primitive cognitive vestiges and new evolutionary additions that together can accomplish amazing things, but in mostly unknown ways. Generally we don't look at the brain that way. We tend to think that the forces of evolution are terrifically efficient at rooting out all wastefulness to make the brain thoroughly optimized for smooth, clean operation. But the truth is that evolution feels its way toward success, tinkering and puttering until it stumbles across marvelously inventive solutions to the problems that the need for survival presents, and then shambles on. Our brain, amazing as it is, is not an efficient machine, but a maddeningly complicated organ that stubbornly resists analysis. We do know a few things, however.

Most adult human brains weigh about three pounds, no matter what the IQ of its owner. It has the consistency of firm gelatin, and scientists currently estimate that the most recently evolved part of the brain—the cerebral cortex—consists of about 30 billion neurons, each synaptically linked to a thousand other neurons.

If we could count all of the particles in the known universe, physicists estimate we would come up with the number 10 followed by 79 zeros. But if you calculate the number of possible neural connections in the brain, you would arrive at the number 10 followed by *1 million* zeros, at least. It would take us 32 million years to tick each one off. If you added up the length of myelinated nerve fibers in the average brain it would come to about 100,000 miles. Most human brains have about 186 million more neurons residing in the left hemisphere rather than the right, but considering the numbers we are dealing with, that difference is really trivial.

But none of these calculations even begins to address the intricate structures and chemistry of the synapses, dendrites, and axons that link all of these cells to one another, or the complex interactions that take place each moment within each of the 50 or so varieties of neurons that perform specialized functions in the brain. In general, though, what neurons do best is store and move information. Each is a masterful engineering accomplishment in itself. A single neuron, for example, holds an average of 1 million sodium pumps in a space a fraction of the meager territory outlined by the period at the end of this sentence. Altogether, that is 100 billion trillion sodium pumps per brain. If they didn't exist, we couldn't think a thought or feel a sensation because they make it possible for each brain cell to receive and pass along impulses to the other neurons it is connected to.

The brain generates its electrical impulses by swapping ions between atoms like bargainers at a flea market. Add ions here and subtract them there, and the atoms they are associated with develop positive or negative charges. When you move a muscle or see a light, neurons predisposed to react to those particular sensations activate ion channels inside themselves. Sodium is pumped through the conduits, and this positively charges a membrane in the neuron. If the sodium channels are highly activated, the electrical signal reaches a "threshold potential" and triggers a nerve impulse, something like the way a loud, insistent hotel customer gets the attention of the bell captain. The impulse is then passed on and travels to its next neuronal destination, eventually gathering up other impulses to generate a thought, a moved hand, or a crippling fear.

But if the sensation doesn't result in enough sodium to pass the impulse along, the neuron remains at rest, unaroused, and diffuses whatever sodium has been pumped its way. The sensation stops there. We don't really know how many flags like this our senses wave at our brain because we aren't aware of all of them. The brain filters them out, or more accurately, their weaknesses filter the sensations themselves out. Good thing, too. Otherwise we would all suffer from a monumental case of attention deficit disorder, incessantly bombarded by every sound, sight, smell, taste, feeling, and thought we experience.

Complex as the individual machinery of each neuron is, the truly defining characteristic of brain cells is their outgoing nature. They live (and die) to be in touch with one another. As you read these words, each neuron in your head is exchanging information at the rate of 200 million operations per second. This, by the standards of the average desktop computer, is ponderous, but what neurons lack in speed they make up for in affability. On average, each of our 100 billion neurons reaches out to 1,000 others, often by long and rambling routes, which keep every sector of the brain in close touch.[5]

This neuronal need to communicate turns out to be one of the reasons why our brains are so large. Though you could be forgiven for thinking that we have grown cerebrally overweight because our brains are packed cheek to jowl with brain cells—at least compared with other mammals—you would be wrong. A rat's cortex, for example, is jammed with 100,000 neurons per cubic millimeter, whereas ours has as little as a tenth as many.[6] But this shortage of neurons doesn't make our brains simpler, it actually makes them more complex. The reason why our neurons are not as densely packed is because they need more elbow room so their dendrites and axons can reach out and connect with other neurons (axons send impulses, dendrites receive them).

The brain cells of a rat and a human are virtually identical— we pretty much do our cerebral organizing with the same cellular equipment. But if you were to remove one neuron from a rat's brain and one from a human *with* all of its connections intact, and then roll the two into separate balls, the human ball would be ten times larger than the rat's. Apparently in the human brain, as in the real world, it is all about whom you know. Amazingly, to conduct all of these chemical and electrical conversations, the brain burns about the same energy each day that a 20-watt lightbulb does.

This need neurons have to be in constant touch means that everything we do and every event we encounter is experienced as one great stream of consciousness. We have a sense that there is someone—ourselves, to be specific—who is seamlessly absorbing the world around us, and the world within us, rather than simply processing random sensations that have no connection. The source of our unique ability to be self-aware has very likely emerged from the rich chattering that our interconnected neurons constantly do. Just as a beehive emerges from the interactions of thousands of bees, or cities arise from the common needs of interacting humans and their environment, our humanness has materialized from the dense neuronal babbling of our brains.

The shuffling and patterning of symbols, the subordination of one concept or memory in favor of another, and the ability to prioritize, says Deacon, are all crucial to language and, ultimately, speech because language is all about assembling information in organized, hierarchical ways. It is evident in nearly every sentence we utter.

Linguists call this shifting and subordination process recursion. Recursion is the way we fold one concept, when we speak, inside another. It reflects the way our mind organizes symbols. And it is the source of the sorts of grammatical conundrums that drove us all crazy in middle school English class when we struggled to unravel concepts like prepositional phrases, dependent clauses, and participles. You can find simple examples in the sentences "She walked behind the building," or "I think he thought of a great idea," or "He realized George thought of a great idea in the morning behind the building."

Experimental psychologist David Premack once imagined two of our ancestors squatting alongside a fire with one of them saying to the other, "Beware of the short beast whose front hoof Bob cracked when, having forgotten his own spear back at camp, he got in a glancing blow with the dull spear he borrowed from Jack."

Premack was actually using this as an example of language being "an embarrassment for evolutionary theory" because he could never imagine why evolution would have created such a complex capability. But cognitive scientist Steven Pinker remarked that Premack's complaint reminded him of the Yiddish expression "What's the matter? Is the bride too beautiful?"

Pinker argues that recursion is absolutely crucial to language and the unique ways in which humans think and express themselves. He says there can be no debate that it is built into the human brain. Communication doesn't have to be tortuous and convoluted, as in Premack's example; it can simply transmit information in a precise pattern and order that clarifies what we mean. At the very least it would have been a tool that vastly improved the chances of survival for any creature who happened to be endowed with it.

"It [recursion] makes the difference whether a far-off region is reached by taking the trail that is in front of the larger tree or the trail that the large tree is in front of," says Pinker. "It makes a difference whether that region has animals that you can eat or animals that can eat you. It makes a difference whether it has fruit that is ripe or fruit that was ripe or fruit that will be ripe. It makes a difference whether you get there if you walk for three days or whether you can get there and walk for three days."[7]

There are other simple forms of recursion that make the point. "Joe, who is very angry, will see you now" not only tells you Joe will see you now, which is the main thought, but also explains the state of mind he is in, which, even though secondary in this sentence, nevertheless provides crucial information. Of course, you could also turn the sentences around. "Joe is angry even though he is going to see you now." The hierarchy of the expressed ideas conveys very different information about Joe's state of mind and, most importantly, its effect on you.

The precise expression of the thoughts we shuffle in our minds has other uses beyond giving clear directions, useful as they are. Where language proves its real worth is in the way we use it with one another because it is that interaction that shapes our relationships. In this may lie the secret of its power, and ultimately the reason why it seemed to blossom out of nowhere.

. . .

An odd thing about our ability to create, shift, and prioritize symbols is that it requires that we be self-aware. Why? The rise of language goes hand-in-hand with the rise of consciousness. The terms "self-aware" and "conscious" are so normal they tend to be almost meaningless to us because we take this state of mind for granted. But we shouldn't, because it is rare outside of our species and it is essential to being human. It requires that we perceive that we have a "self" that is distinct from others and the rest of the world. If we could not make this distinction, we could not purposefully make a tool or get up off of a chair and walk across a room to shake someone's hand because we would not truly be aware that we are different from the chair or the room or the person across the room. Nor could we consciously manipulate thoughts the way we manipulate objects, because to physically and mentally manipulate anything—on purpose—requires us to be aware that there is the manipulated and the manipulator.

Our *self*-awareness begins in a very concrete way with the knowledge that we all have bodies—the most direct indication that we exist. On the most basic level this is the way we draw the line that separates our selves and the rest of the world. Neurologist Oliver Sacks makes this point about as well as it can be made with a story about one of his earliest experiences as a medical student. One evening a nurse called him to a hospital room where he found a patient lying on the floor beside his bed, gazing appalled and disgusted at his own leg. Sacks asked the young man if he could help him get back into bed, but he kept shaking his head no as he continued to stare horrified at his leg.

There was nothing wrong with the leg so far as Sacks could see, and he couldn't understand why the man was so upset, so he asked him what was the matter. The patient told him he had been admitted to the hospital earlier in the day for tests because his left leg felt "lazy." At about dusk he fell asleep and then later woke up to find, to his horror and disgust, a severed leg in his bed with him! He couldn't imagine where the thing had come from. After inspecting it, he told Sacks, he picked the leg up and struggled to throw it out of the bed, but somehow when he did he had gone with it, and now, as he sat on the floor, the limb had become attached to him.

"Look at it!" he yelled at Sacks. "Have you ever seen such a creepy, horrible thing?" He then seized it with both hands and tried with all his might to tear it off. When that didn't work, he began to punch it. Sacks, who was now squatting beside the man on the floor, suggested he not do that.

"And why not?" asked the man.

"Because it's your leg."

This absolutely stunned the patient. He couldn't believe it. He couldn't because he was thoroughly convinced the leg was not attached to him. So finally Sacks asked him, "If this—this thing—is not your left leg, then where is your own left leg?"

The man thought about that. "I don't know," he finally answered. "I have no idea. It's disappeared. It's gone. It's nowhere to be found."[8]

Amazing as this story is, it is true that people do sometimes lose track of themselves or a sense of themselves or a part of themselves. Somehow the lines between them and the rest of the world disappear. In this case it turns out the patient was suffering from a disease known as Pötzl's syndrome or optic-kinesthetic allesthesia. It is a form of dissociation from one's self, a cousin to diseases like multiple personalities syndrome, in which people experience living as several distinct people or personalities, all of them inside the same body. Damage to the posterior portions of the right hemisphere of this man's brain, which very specifically controls awareness, or gnosis, of the left leg, had slipped its moorings and caused him to literally lose a sense of himself, or at least that particular part of himself.

Damage like this can be caused in several ways—stroke, concussion, brain tumor—any number of afflictions. Depending on the side of the brain that is damaged, Pötzl's syndrome can affect your eyes, speech, or limbs. It could happen to any of us at any time, and it powerfully illustrates that our sense that our bodies and our "selves" are one and the same is manufactured in our brain, along with the rest of our reality. If the brain is physically damaged, our sense of self can warp or splinter or vanish. Injuries such as these reveal that the "self" is a very fragile thing that depends on how certain clusters of neurons connect, exchange hormones, and fire.[9]

Keep this in mind the next morning you groggily pick up a cup of coffee and bring it to your lips. Ask yourself how you know it is you who is picking the cup up. Why don't you think it is another hand that belongs to another person?

This question isn't as insane as it sounds. The simple act of drinking coffee requires that you not only have the physical ability to reach out, grasp the cup, and bring it to your mouth, it also requires that you perceive a "self" that is directing and following through on all of those actions. For you to succeed in drinking, you must have the perception, on some deep level, that you are separate and different from the world around you, and that the "you" drinking the coffee is all one piece, a unified whole, not splintered and separated. This makes you "self"-aware. It means, in fact, that upon reflection, you understand something amazing: that someone you call "you" exists.

All of this seems obvious to most of us most of the time. "I am me," you say. "How else can it be?" Except, as much as we take this ability to refer to our "self" for granted, in nature it is actually extremely rare, as psychologist Gordon Gallup proved in a famous series of experiments in the 1970s and '80s.

Gallup wondered if other primates besides us might be self-aware, at least in some sense. To get to the bottom of the question he conceived of a clever experiment. He anaesthetized different types of primates, including orangutans, monkeys, chimps, and gorillas, and placed an odorless but clearly visible mark on their foreheads. When they awoke, each animal was placed in front of a mirror. If it looked in the mirror and saw the mark and then touched or rubbed the mark, as opposed to reaching out and touching the mark in the image in the mirror, Gallup concluded that they understood they were seeing a reflection of themselves, as opposed to an entirely different animal. In other words, they were, on some level, self-aware.

Monkeys failed the test. Orangutans, after a little time, didn't. Chimps never failed and, surprisingly, gorillas mostly did (though some scientists have argued that there may be other reasons for this than a lack of simian self-awareness).[10]

Having an image of your own body *as* your body is a prerequisite for the special brand of self-consciousness we all blithely live with each day. It means that unlike birds and possums, the green skeleton frogs of Madagascar, or my dog Jack, you are not unconsciously moving through the world, instinctively reacting to the forces around you. You are aware that you are doing these things, and you are actively and purposefully taking some control.

Without that awareness, you would be no better off than the monkey that looks in the mirror and treats its own reflection as a total stranger. Or, in some way, you would be like the man who was unable to recognize his own leg for something other than the severed limb of a cadaver. Somehow you would miss the point that your body and you are one and the same. Your self and your environment would blend into one another.

. . .

A prerequisite of being conscious of your self is that you be conscious, period. And that raises still another perplexing mystery: How do you explain a mass of one hundred billion neurons the consistency of Jell-O weighing roughly three and a half pounds conjuring up something as miraculous as a human mind?

After winning the Nobel Prize in 1972 for insights into the chemical intricacies of the immune system, Gerald Edelman began to tackle precisely that question.[11] In 1981 he founded The Neurosciences Institute, where he assembled scores of scientists in fields from biochemistry to artificial intelligence and neuroanatomy to explore how the physical interactions of the human brain make consciousness possible.

Edelman concluded that our brains are so deeply interconnected and have so thoroughly woven the primal parts of themselves into their more recent cerebral additions—the prefrontal cortex, for example—that consciousness emerges out of the trillions of interactions that take place at any given moment. It's strange to think of something as singularly crucial as consciousness as an evolutionary by-product, but if Edelman is right, that's where the evidence points.

Neuroanatomists divide the brain into six broad sectors: the thalamus, brain stem, cerebral cortex, basal ganglia, hippocampus, and cerebellum. Edelman holds that the brain's internal diversity and the way those diverse parts, both ancient and newly evolved, interact creates our special brand of human consciousness. A central player in Edelman's theory of consciousness is the thalamus, which is, for want of a better term, the brain's sensory gatekeeper. It sits—gray, oval-shaped, and dense with neurons—midway between the brain stem (which attaches the brain through your neck to the spinal cord) and the prefrontal cortex. Nothing we experience—the touch of a hand, a bright light, a smell, or the passing of a cool breeze—makes its way to the cerebral cortex without first passing through the thalamus. Its dense network of looping connections called the thalamocortical system constitutes a "meshwork" of communication lines that reach into every area of the brain.

Without this meshwork, Edelman believes there would be no consciousness because consciousness requires that the brain be continually checking in with every far-flung sector of itself. And that is what the thalamocortical system excels at. It pulls and pools information from the basal ganglia, for example, which handle the planning and execution of complex motor skills; or the hippocampus, which specializes in shifting important short-term memories into longer-term storage; or the cerebellum, in the back of the brain, which helps coordinate and synchronize motion (but—according to recent discoveries—is also important to speech). In addition to these, the brain houses millions of clusters of neurons that perform specific functions, such as dealing with loud sounds versus soft ones, or linking smells with their origins in space.

Each of these clusters "summarizes" or modulates the information that cascades around the brain the way the vibrating strings of a guitar modulate and interact with one another to create the sound of a strummed chord. This "summarized" signal is then returned to the feedback system, where it mixes with and changes other incoming summarized signals. The modifications never stop. Edelman likens these larger interactions to the interplay among the members of a string quartet.[12] He writes,

"Imagine a peculiar (even weird) string quartet in which each player responds by improvisation to the ideas and cues of his or her own, as well as to all of the kinds of sensory cues in the environment. Since there is no score, each player would provide his or her own characteristic tunes, but initially these tunes would not be coordinated with those of the other players. Now imagine that the bodies of the players are connected to each other by myriad, fine threads so that their actions and movements are rapidly conveyed back and forth through the signals of changing thread tensions that act simultaneously to time each player's actions. Signals that instantaneously connect the four players would lead to a correlation of their sounds; thus more, new, cohesive, and more integrated sounds would emerge out of the otherwise independent efforts of each player. This correlative process would also alter the next action of each player, and by these means the process would be repeated but with new emergent tunes that were even more correlated. Although no conductor would instruct or coordinate the group and each player would still maintain his or her style and role, the players' overall production would tend to be more integrated and more coordinated, and such integration would lead to a kind of mutually coherent music that each one acting alone could not produce."[13]

Edelman calls this synaptic dance "reentry" because it requires that all of the interacting information throughout the thalamocortical system keeps looping back, exiting, and then reentering. A simpler (but by no means simple) version of this happens when we look around. Research into the workings of the visual cortex—one of the better-understood areas of the brain—has revealed that separate parts of the cortex process the sensations of color, shape, and movement. None of them is managed by a central "boss," any more than Edelman's string quartet is guided by a conductor, or even sheets of music. Instead the visual information comes into the brain's meshwork, interacts, and in the process of reentry creates a literal picture of the world, all of it formed by integrating disparate pieces of information, in the blink of an eye. What is more, these images appear whole and continuous even though different parts of the brain didn't initially experience them that way. The visual cortex takes all of the signals and melds them into one seamless series of events.

Consciousness, Edelman says, happens in a similar way. The brain handles information of all kinds, both internal and external. The raw information is modulated in areas that manage sight, sound, touch, and every other sensation. In a snap, long- and short-term memory are also consulted to see if the sensation is familiar. Meanwhile, neuromodulators that are squirted throughout the brain by the norepinephrine system to encode sensations as pleasing or terrifying or disgusting, add additional information.

But only some experiences emerge into consciousness because still another selective, Darwinian-like process is at work.[14] If the same information keeps reentering the system, then it has "survived" and is selected. It is nudging, sometimes shoving, the cerebral cortex, telling it that this information is an accurate representation of an experience and it should be paid attention to. The wind really *must* be blowing, that really *is* the scent of a predator, an attractive member of the opposite sex *does* want to mate. I am scared. I am elated. I am hurt. The strength and the timing of the information are related to how insistently it reenters the system, and the more insistently it reenters, the more likely it will emerge into conscious thought.

The emergent nature of consciousness makes it the most extreme and dramatic example of what emergent systems can create. If we look back over evolution, all forms of life and all natural systems seemingly evolve out of behavior that looks chaotic, at least, insofar as they are never designed and created from the top down, not even when the result is something as remarkable as a self-conscious, thinking, talking primate. But the thing that makes self-consciousness really different is that for the first time it is a behavior that is aware it is behaving. This is something profoundly new in nature.

If Gallup's tests are any indications, self-consciousness in our ancestors arose by degree. Earlier hominids did not "abstract" in the way we do. They very likely shifted their attention and changed their behavior in reaction to their environment and conditioning the way most other primates do. If they were cold, they would try to get warm. If they heard a growl or saw a sudden movement, they turned to fight or turned tail in flight.

But as our predecessors grew more self-aware, they would have begun to stop simply reacting to their environment and begun to purposefully shift their attention and take greater control of it. Why was the line of primates that led to the human race capable of this? Perhaps because earlier in their evolution their thumbs and hands enabled them to make tools, which required them to manipulate objects, which, in turn, required putting one thing aside while another thing was taken care of. And in time their increasingly sophisticated brains enabled them to apply the same talent to virtual, imaginary objects that existed only within their minds. But the wetware that ultimately pushed them into a being that was human was the prefrontal cortex.

This part of the brain, says Edelman, made the difference between something he calls "primary consciousness"—the sort of awareness of the world that many mammals experience—and "higher consciousness," that special brand of self- and world-awareness that you and I experience every waking moment. If chimps and orangutans today have some sense of self, and if toolmaking and gestural language require some level of self-awareness, then *Homo erectus* probably lived life in a twilight zone somewhere between chimps and us, more self-aware than any other creature of its day, but far from deeply thoughtful.

The question is what other forces were at work that built brains that became capable of something as remarkable as self-awareness? One of the key influences, maybe *the* key influence, was not simply the physical world, not even, strictly speaking, our inner, thought world, but the place where we interact with one another: our social world—the one space where all of our life's key forces converge.

Chapter 7

Words, Grooming, and the Opposite Sex

*Once you have become permanently startled, as I am, by the realization
that we are a social species, you tend to keep an eye out for pieces of
evidence that this is, by and large, a good thing for us.*
—Lewis Thomas

HUMANS ARE THE MOST SOCIAL CREATURES on Earth. Our
need to interrelate is deeply threaded throughout our DNA and our
brains. Chimps and gorillas are also extremely social, which means the common ancestor we share with them very likely was as well. Jane Goodall de-
scribes two chimp friends she watched when they were reunited after being
separated in Gombe National Park. At first sight both chimps hopped and
screamed and danced and hugged at being together again. They acted the
way you and I might after seeing a long-lost relative. But then they did
something we probably wouldn't. They began to groom one another. They
huddled together and attentively and affectionately set to work picking nits.

There are good reasons for doing this in the wild. The jungles of Africa
are filled with fleas, ticks, and parasites that survive by hopping a ride on
the other living creatures around them. If left to their own devices, they
would drain their hosts dry or infest them with enough disease to kill them
before either they or their hosts could reproduce. So grooming serves a real

purpose; it is a survival technique, and having two or more sets of hands to do the work is clearly better than one.

But there may be other reasons grooming evolved. It is also a way of connecting. When chimps comb through one another's fur, calming neurotransmitters cascade into their brains that make them feel warm and safe.[1] You and I experience the same soothing echoes of grooming when someone we care for runs his or her fingers through our hair, strokes our arm, or holds our hand.

Our tree-swinging ancestors had very good reasons to connect as strongly with one another as possible, given that the environment they lived in was filled with danger. Psychologists Robert Seyfarth and Dorothy Cheney have found that in the jungle, grooming relationships can improve a primate's chances of staying alive, accumulating power, building successful alliances, and making babies. In the early 1990s both scientists[2] conducted studies of vervet monkeys and found that they will pay much more attention to the distress calls of individual monkeys they have recently groomed than those they haven't. In fact, the more attentive the grooming, the more likely different individuals were to help one another out.

This makes grooming both a survival technique and a form of communication. In fact, Liverpudlian psychologist Robin Dunbar has speculated that nitpicking may have laid the early foundations for human conversation, the simian version of a phone call, or a chat over a drink at the local bar.* Earlier primates may have started out staying in touch by tidying one another up, he says, but in time we expanded that primal form of socialization and communication to include a new and remarkably potent talent we were developing: speech. Following Dunbar's logic, speech must then be about more than the practicalities of exchanging information. It also must involve vulnerability and emotion, not to mention scheming and manipulation.

. . .

Seeing similarities between the relationships we develop in the human world and those in the simian one isn't new. Goodall watched plenty of complex, human-style dramas unfold among the cast of primate characters

* The original meaning of the English word "chat" comes from the chattering sound made when monkeys (and later humans) disposed of parasites with their teeth as they groomed others. John Skoyles and Dorion Sagan, *Up from Dragons* (New York: McGraw-Hill, 2002), p. 83.

she watched and named during her years in Gombe. It is easy to see ourselves in the alpha chimp Goliath's fall from power, Merlin's slow march to madness, Flo's unlikely attractiveness and never-ending romances, the regal warmth and wisdom of David Graybeard, and the touching friendship between David and William, a story that could have inspired *Of Mice and Men* or *Midnight Cowboy*.[3]

After years of scrutinizing the likes of Flint and Figan, Gilka and Goblin, Goodall also made some disturbing and surprising discoveries that remind us of ourselves. She found that chimpanzees are capable of murder, organized hunting, complex relationships, and convoluted emotions. Every one of these discoveries astounded the scientific world, and revealed how the smoldering embers of our own humanity may have found their way across time into the world we inhabit today. Change the setting, push the complexity of the relationships, the deceptions, and the struggles for status, and the personal losses and victories begin to look like the plots of Austen, Trollope, or Hemingway.

If Goodall's stories have a human ring, you can imagine that by the time *Homo erectus* had arrived, the dramas unfolding among these troops of migrating toolmakers and hunters were beginning to look even more familiar. After all, they had placed a good deal of distance between themselves and the primordial rain forests of their tree-swinging ancestors. They were now fully bipedal, their hands were free, and their thumbs were entirely opposed. Their brains had more than doubled in size since Lucy's time and were still growing rapidly. Erogenous zones for some time now had been sending clear sexual signals, while at the same time their faces were becoming more expressive, conveying in greater depth the complicated thought-world evolving within. And narrowed birth canals forced the arrival of "younger" children that were more helpless and, above all, more impressionable than ever. They were growing up more intelligent than any other earthly creature. All of this made them increasingly less confined by the dictates of their genes.

The day-to-day social relationships within the troop would also have become increasingly complex. Each member had to compete, subtly or otherwise, not simply for mates with the best DNA, but also for mates that would be reliable caregivers and partners. Primal versions of fidelity and truth, not simply sex and brute strength, had become key forces in the battle for survival. (See chapter 2.) Cooperation was paramount, but so were social talents like shrewdness and insight. The growing complexities of sexual and social politics presented a whole new kind of evolutionary challenge that upped the social stakes.

Handling all of this intrigue required sophisticated communication and big brains capable of powering it. But which came first? The need or the brains? Speech that smoothed communication or cerebral adaptations that enabled refined forms of it? For decades the common wisdom among anthropologists has been that the demands of hunting in groups, making tools and weapons, and eating more meat were the primary drivers of brain growth. And all of those almost certainly contributed to our intellectual evolution.

But could they alone have been responsible for the wealth of neurons we all enjoy today? They leave out one immensely important part of our ancestors' daily life: their struggle to deal with one another, to compete for power, for status, for the affections of potential mates while, at the very same time, building close alliances and friendships. Is it possible that complex social dynamics like these were the true primary movers pushing the extraordinary development of our intellects?

. . .

In 1988, two British psychologists, Richard Byrne and Andrew Whiten,[4] proposed that monkeys and apes often observed the behavior of other monkeys and apes as they interacted, and then used that knowledge to decide how *they* would act toward those they had watched. If they noticed that an ape was aggressive, for example, then they tended to be more deferential when dealing with that ape. Or if they noticed a particular ape was generous, they might try, at some point, to take advantage of that generosity. In effect Whiten and Byrne found that primates, simply in the course of their daily lives, used the knowledge they gathered interacting with one another to manipulate individuals in a social group to their own advantage. This meant they were juggling an enormous amount of information almost constantly. Byrne and Whiten called this the Machiavellian Intelligence Hypothesis.[5]

If apes do this, then our ancestors surely did as well—at least that was the thought Robin Dunbar had in mind when, in the mid-1990s, he began to explore the work of Byrne and Whiten. Dunbar theorized that the more intelligent our ancestors became, the more complex their increased intelligence made their lives. And the more complicated their lives, the more intelligence they needed to handle it all. In fact, he found a direct correlation between the number of relationships individual primates deal with in a group and the size of their brains, or more precisely their neocortices.

The neocortex is the "thinking part" of any mammal's brain, as opposed to the limbic system or the brain stem, both of which are considerably more ancient. In most mammals the neocortex comprises about 30 to 40 percent of total brain mass. In some primates it can claim up to 50 percent of the cerebral real estate. But in humans it accounts for a whopping 80 percent, an area that packs roughly 100 million yards of axons and dendrites into a space the size and thickness of a formal dinner napkin. Its various regions handle all of the higher levels of cognition—planning, imagining, language, spatial calculations, and the intelligent consideration of what we see, feel, hear, smell, and touch.* And most of it evolved in the short evolutionary space of the past one million years.

Dunbar's studies revealed that when the size of a group of primates increased by one, the number of relationships a member of that troop now had to track increased by one *plus* all of the connected relationships that both members had in common.

Consider what would happen if a primate named Joe lived in the same troop with Mike. And imagine that Joe was a friend of Mary's, but Mike didn't know Mary. Under those circumstances Joe and Mike might normally compete, sometimes ferociously, for Mary's attention. *But* let's say Mike *did* know Mary, and let's further assume they were allies. Since Joe wouldn't normally want to anger Mary (they are friends), he might be a touch more cordial to Mike (unless, of course, Mary was his mate and Mike was trying to horn in, but that's another story).

If this sounds complicated, that's because it is. One-on-one relationships are a two-way street. Add third and fourth parties and you get a traffic jam. The complexity of any social interaction goes up exponentially every time you add another to the mix. Dunbar calculated that within a group of 20 primates you would have to keep track of 19 direct relationships, but 171 indirect relations that those 19 have with others.[6] So if you go from, say,

---

* The prefrontal cortex is part of the neocortex but is not the same as the neocortex. While current theory holds that much of the human neocortex evolved over the past one million years, scientists believe the prefrontal cortex emerged in just the past four hundred thousand, or less.

4 direct, close relationships to 20, then your immediate circle of friends increases fivefold, but the indirect, ever-changing relationships you now need to constantly keep in mind increase thirtyfold!

This is even more complicated than it first appears because not only does the number of relationships increase, but because they are all intertwined and constantly changing, the emotional complexity rises as well. For every action within the troop, there isn't one, simple opposite and equal reaction, there were multiple opposite and unequal reactions that spread like wild rows of falling dominoes. None of it would have had the clean feel of Newtonian physics or algebraic equations. It would have had the feel of chaos, and chaos—ever-changing, unpredictable situations—requires as much intelligence as a creature can muster. (As a species, we abhor uncertainty, and so our brains constantly struggle to reduce it.) Think of how much time and energy we put into imagining what the boss is thinking or how we can outmaneuver a rival, plant an idea, win the affection of a lover, or consider the best way to reward or discipline our children.

What made the world of our ancestors peculiarly tough was that they could not afford to make enemies within the troop any more than we can within the circle of people we deal with in our daily lives. Their world was too small. If every conflict among these troops came to a fight to the death, we would have evolved into nothing more than a species of vicious and conniving primates. Or more likely we would be extinct. But mostly we aren't vicious and conniving (though we certainly can be). Mostly we are wary, but open. It is more in our nature to trust than fear. That means that winning friends rather than bullying must have evolved as an essential survival technique.

Given this situation, a new tool was required. Not a flint knife or a hand ax, but a talent for understanding others and helping them to understand you. A talent for communication.

These circumstances would have created a kind of arms race: our ancestors' brains would have evolved greater intelligence to handle the increasing complexity of intratroop relationships, which would have in turn selected for increasingly intelligent minds that could track the progressively complicated relationships, and so on. Though Dunbar's work hints at this arms race and it makes intuitive sense, a genetic study published in 2005 and conducted at Howard Hughes Medical Institute indicates that the genes that create the instructions for building the human brain did indeed change enormously, especially in the past four hundred thousand years. The study, conducted by a team headed by geneticist Bruce Lahn, at the University of Chicago, compared the sequence of the human Abnormal Spindle-Like Microcephaly Associated (ASPM) gene to the same gene in six other primate species—chimpanzees, gorillas, orangutans, gibbons, macaques, and owl monkeys (as well as cows, sheep, cats, dogs, mice, and rats). ASPM genes are strings of DNA linked to a severe reduction in the size of the cerebral cortex, parts of which handle planning, abstract reasoning, and other higher brain functions. The analysis revealed that the human brain's genes had changed markedly, apparently to allow increased brain growth, whereas the other primates' hadn't.

The forces driving these rapid changes had to be powerful; otherwise we wouldn't see them in the genetic record. Like Dunbar, Lahn suspects it was the social pressures created by increasingly intelligent and complex interactions that drove the selection for smarter hominids. After all, said Lahn, "As humans become more social, differences in intelligence will translate into much greater differences in fitness, because you can manipulate your social structure to your advantage." Put another way, smooth social skills must pay off handsomely.

. . .

If smooth social skills are an advantage, and if they require equally smooth communication, does this explain the evolution of speech? Why not simply evolve into highly intelligent creatures who communicated by lovingly picking parasites out of one another's bodies? Dunbar's answer is that when it comes to communication, grooming, soothing and useful as it was, had its limitations. Not only was it vague as a way to communicate, but it also was a one-on-one activity. You can't groom two or three or four others simultaneously, but you can hold a conversation with more than one. Eventually, as brain size and troop size grew, and as more communication was needed, grooming simply failed to allow enough time for all of the members of the troop to "converse" and to keep track of one another.

Dunbar theorizes that this shift would have first begun about two million years ago, around the time *Homo habilis* emerged. Shortly afterward, the first rudimentary forms of speech would have followed. Forced into a corner of sorts, evolution would have stumbled upon a more efficient way of staying in touch, not with hands and fingers, not even by gesturing, but by grooming, as Dunbar puts it, "at a distance"—making sounds at one another that accomplished the same social, if not hygienic, goal. The big advantage of "sound" grooming would have been that our ancestors could more effectively track all of the complex and far-flung relationships within the troop.[7]

Dunbar doesn't subscribe to the gesture theory of language origin. For him, language finds its foundations in the contact and alarm calls of African monkeys and apes—the sounds they automatically scream and hoot when they are in danger or want to get the attention of those around them. He bases this belief on the work of Seyfarth and Cheney. Their exhaustive observations of vervet monkeys has revealed that they use different calls to warn of predators: one for eagles, another for snakes, still another for leopards, and so on—a simple kind of vocabulary where a sound stands for, or symbolizes, different dangers.[8] Gelada baboons use subtly different moans and whinnies and grunts to communicate simple messages when grooming. Something like these, he believes, are sounds that in our ancestral past could have represented meanings that evolved into the stabs at some of the first words. (This is generally reminiscent of the Pooh-Pooh theory of language.)

But as the time budgets for grooming were squeezed, vocal chatter would have had to increase to supplant less efficient one-on-one grooming. Dunbar estimates that between two million and five hundred thousand years ago, creatures who were roaming in increasingly larger troops would have been forced to spend about 30 percent of their time grooming and communicating if they hoped to successfully keep tabs on one another.[9] Dunbar believes that this would have merged with larger brains and greater vocal control to create the first protospeech.[10]

What Dunbar's theory doesn't directly address is the issue of mentally manipulating symbols, something that gesture and signing deal with so well. But maybe there is a middle ground where both Dunbar's theories and theories of gesture and symbol manipulation can meet. Maybe the whines and moans and grunts of grooming are more closely related to prosody, the meanings we bring to our conversation in the tone and inflection of voices, whereas the words themselves, their meaning, syntax, and organization, are more closely related to the symbol shuffling that hands and gesture and, ultimately, Broca's and Wernicke's areas excel at. Both emotion and content are crucial to language as we know it; without both, any kind of communication would be crippled.

Like Broca, Karl Wernicke, a Polish-born neurologist and psychiatrist trained in Germany, discovered the part of the brain named for him in 1874 by studying those who suffered damage to a specific cerebral area that impaired their ability to communicate. Except in Broca's case, the impairments were related to the ability to process and understand speech (or signed language), not to generate it.

Wernicke's area is in the temporal lobe on the left side of the brain (in most humans), where it meets the parietal lobe, roughly behind the left ear next to the primary auditory cortex. Wernicke's area is linked to Broca's area. Without Wernicke's area we would be incapable of understanding any language we hear as language. Those who suffer from Wernicke's aphasia (or receptive aphasia) are unable to comprehend what others are saying to them.

Brain scans indicate that Wernicke's area plays a role in processing word sounds as speech and then searches the mind's dictionary before passing their meanings along to other parts of the brain.

Wernicke's area is also often affected in those who suffer from schizophrenia and may explain why they sometimes hear hallucinatory "voices" that talk to them and that feel completely real.

The discoveries of parts of the brain like Broca's and Wernicke's areas indicate that certain abilities seem to reside in extremely specific locations. For example, neurologists have found one tiny area of the brain—about 1 centimeter square—that activates only when consonants are heard! Some Wernicke aphasics are afflicted with an impairment called anomia, which makes it impossible for them to name certain objects, sometimes very particular ones—body parts or vehicles, colors or proper names, for example. One patient could not name fruits or vegetables, which prompted psychologist Edgar Zurif to joke that the syndrome should be called "banananomia."[11] But oftentimes that isn't the case, and it seems impossible to locate any one region that owns a particular function.

At the same time, both grooming and symbol manipulation touch directly upon two other equally crucial aspects of language: the ways we use it to create and manage our personal relationships, and the ways we converse with ourselves to shape our own mental and emotional lives—a kind of self-grooming, otherwise known as consciousness.

These two parts of us—our social selves and our inner selves—are inextricably bound. In fact, over the course of our lives, they define and remake one another, and their fusion is what has made the cultural progress that differentiates us so thoroughly from all other creatures.

Language is the all-purpose tool we have used to create human culture. But we could never have tethered our minds together to realize this worldbuilding if we hadn't first managed to cooperate and connect with one another. That's why the emotional aspects of language are so central. In the end, nothing would get done if we couldn't use language and the mind it has created, to adhere somehow to one another and get along.

Connecting emotionally is so important, in fact, that most of us spend the majority of our time tuning into the apparently "impractical" aspects of communication and conversation. Fully two-thirds (on average) of the talking we do is not at all pragmatic in the sense that it accomplishes anything we might call work.[12] Mostly we don't talk about how we are going to complete this project or that, get the car fixed, or find our way from point A to point B. Instead, we jabber on about personal experiences, likes and dislikes, and the ways our relationships are going well or in circles or down the tubes. This is the grooming aspect of conversation—just touching, gossiping, checking in, and sharing.

We know this from another experiment that Robin Dunbar and his students performed. They eavesdropped on hundreds of conversations and monitored what each person was talking about several times a minute. The practicalities of work, religion, politics, even sports usually monopolized no more than 10 percent of the conversations. The conclusion: All of this talking about others, and by implication our relationships or opinions about them, allows us to keep track of and influence others' thinking about the friends, competitors, lovers, office workers, and family that run in our common circle, our troop. We are not simply tracking everyone else, says Dunbar, we are trying to synchronize what the troop thinks about us with what we *want* the troop to think about us. We are trying to match our self-image with our social image.

Think of conversations among teenagers. They act as complex feedback loops that each teen uses to project an inner image to the rest of the group in the hope that that image will be accepted. If it is, then that acceptance in turn reinforces the teen's inner image. Each party jockeys for position, advertising the traits that make him or her an important member of the group, which is to say datable, influential, worth listening to. Teens might use charm, humor, sex appeal, kindness, creativity, even bullying to accomplish the goal, but the work is unrelenting and always evident. They are learning to be both individuals and social animals at the same time. As we grow older, the methods may become more subtle and refined, but they never stop.[13]

The whole issue of successfully managing our reputations is important in all company, but it is particularly important when members of the opposite sex are around. Another Dunbar study reveals that when a group is all male, the discussion of ethics, business, and religion consumes no more than 5 percent of the conversation. But when women and men are both involved, it rises to 15 to 20 percent. Dunbar calls this "vocal lekking." Lekking is a behavior in the animal kingdom that males of the species use to show off what they have to offer potential mates. Dunbar found that younger men in a group spend two-thirds of their time talking about themselves. In other words, they are strutting their stuff—verbally.

Apparently this works. Anthropologists have found that tribal chiefs are often both gifted orators and highly polygenous, which means they excel at attracting and keeping multiple mates, a practice that very effectively spreads that chief's particular genes into the next generation.[14]

Women, as a group, however, talk considerably less about themselves than men. But this isn't necessarily a sign of weakness or deference. They may be keeping their mouths shut so they can sit back and evaluate what the males in the conversation have to offer, the way peahens decide which peacocks they want to mate with by comparing tail feathers.

These sorts of social minglings, loaded with their constant monitoring, jockeying, imagining, and lekking, could only have accelerated the social complexity versus brain arms race. After all, it is hard cerebral work trying to imagine what a potential mate might find most attractive in you while at the same time keeping an eye on the competition so you can outmatch them. Nothing is more changeable, even chaotic, on a day-to-day basis than the shifting tides of personal relationships. And nothing is more important. It requires constant use of the prefrontal cortex, the continual construction and redesign of strategies, plans, and scenarios, not to mention the rationalizations we routinely use to explain to ourselves why we, or others, do the things we do.

Communication, in the form of a budding language, would have been the primary tool in managing those relationships, and evolution would have favored early humans who used their masterful communication skills to better handle them. Applying language to the organization of the hunt or to explaining how to whack an ax head would certainly have been useful, but in the big scheme of things it wouldn't hold a candle to the advantages that a silver tongue delivered when telling a riveting story about how the mammoth was felled. Eloquence contributes to how others come to form their image of us, and that image is central to our personal, emotional, and social lives. They are all connected because the persona that emerges out of the groupthink of the troop is crucial not only to how others see us but also to how we see ourselves.

This interpersonal feedback loop means that we understand problems more easily when they are connected to social situations; we have the inborn capability to unravel social problems because our brains evolved to deal with them.[15] Linda Cosmides, a scientist at the University of California, Santa Barbara, has even argued that humans have a special ability to recognize instances when social contracts are being violated, or when actions do not sync with words. This is crucial because members of groups will disintegrate if trust erodes, and soon instead of safety in numbers, chaos reigns and devolves into every man for himself.[16,17]

Another way of putting this is to say that our personal behavior is enforced not just from the outside, but from the inside, too. If we fail the troop, we risk becoming isolated, and no one wants to be isolated, companionless, and powerless. So we try very hard to get along with others.

This is one of the great ironies of human nature. We are self-aware enough that we want to express our individuality, but dependent enough on those around us that we also want, sometimes desperately, to fit in. In many ways our lifelong goal is to somehow strike a reasonable balance between the two. In this way the language we use to glue us together also shapes what we tell ourselves is acceptable behavior. This is what John

Skoyles and Dorion Sagan call "the troop within our heads" in their book *Up from Dragons.*

. . .

We have, thanks again to our prefrontal cortex, an ability that psychologists call Theory of Mind (ToM), a talent for guessing at what we think others may be thinking. It is absolutely central to imagining how others may react to what we do. Whether we want to please or deceive them, correctly guessing what is on others' minds improves the chances of us getting what we want.

When Theory of Mind and the troop inside of our head combine, something like a moral sense emerges. If, for example, taking all of the bananas was going to feed one of our ancestors in the short term, but make him a social outcast in the long term, then he may have decided not to take them, or at least have decided to share some of them. Or if sharing all of the bananas was going to make him powerful, he may have decided to be generous rather than selfish. Either way, in imagining what others may think is acceptable behavior *and also* in your own long-term best interest, you will very likely curb your more selfish desires. There is a kind of justice in that.

Anyone who is particularly good at this kind of thinking is going to have an evolutionary advantage and an advantage in life. It's the classic, "I know that he knows that I know he really wants the chocolate ice cream, not the vanilla." It harkens back to game theory and mirror neurons, the ability to put yourself into someone else's shoes. This is the prefrontal cortex laboring overtime. It is also the troop hard at work inside your head, and once it is there (and we all have it, starting with the earliest lessons we learn at the feet of our parents), it is there to stay, shaping the people we are.[18]

All of these forces—social, mental, and emotional—with which we endeavor to control the behavior of the person we call our "self" really constitute the thing we call consciousness. Human self-awareness and conscious, purposeful behavior really can't be separated. But it's not altogether clear how the more ancient, unconscious drives that we carry with us from our ancestral past influence how we act or why we act that way.

Cognitive scientist Michael Gazzaniga has a theory about this. He believes it can all be traced very directly to the place in our brain where speech is being processed.

· · ·

For some reason, in most humans, both the generation and comprehension of language is handled on the left side of the brain. There are various theories as to why. Some believe it may be related to handedness.[19] The left hemisphere of the brain governs control of the right side of the body, including our right hands. Most humans are right-handed. There is some evidence that beneath the skulls of *Homo habilis* there was just the slightest bulge of an incipient Broca's area. Since Broca's area is involved in controlling the hands as well as in generating speech, maybe there is some connection between right-handedness and speech, some specialization that was going on. After all, this part of the brain already housed the computational circuitry necessary for using hands to arrange objects. Maybe it was a natural location for the deliberate, sequential control of symbols and words as well.

In *The Language Instinct* Steven Pinker theorizes that language may have taken up residence in the left side of the brain because it wasn't very concerned with dimensions such as space and direction. "Human language may have been concentrated in one hemisphere," he says, "because it . . . is coordinated in time but not environmental space: words are strung together in order but do not have to be aimed in various directions."[20]

In his book on the origins of language, psychologist Michael Corballis argues that at some point there may have been a single-gene mutation that created a strong bias toward both right-handedness and left-cerebral dominance for language. In short, it was an evolutionary accident, and in another universe, had certain genes mutated differently, we would be mostly left-handed and right-headed.

It is also possible that the right hemisphere was simply crammed to capacity when we first began to develop the neuronal bricks and mortar for language, so the gray matter had to find room in the left hemisphere. The right hemisphere already had long been busy perceiving emotional cues as well as all nonverbal but very powerful forms of communication, such as shapes and facial expressions. Experiments have even shown that a mother carrying her baby tends to detect the soundless grimace of her child if she is holding the baby on her left side, which is controlled by the right hemisphere. This may explain why most mothers tend to hold their babies in their left arm. Even monkeys and apes respond more intensely and more quickly when the facial signals of other monkeys are more visible to the left sides of their faces.[21]

Whatever the reasons why speech and verbal thought are generated in the brain's left hemisphere, Michael Gazzaniga feels it is central to the nature of our species. Over the years his work has focused on the two hemispheres of the brain, and he has conducted exhaustive tests with "split-brain" patients, individuals who have had the thick bundle of 250 million nerves called the corpus collosum that connect the two hemispheres of the brain surgically severed. This is usually a last-resort treatment for rare cases of epilepsy where the seizures make normal life impossible.

For most people their hemispheres work just fine after the operation, and the brain seems to mysteriously find other ways to move information between them. But under certain experimental conditions it becomes clear that in split-brain patients the two sides of the brain are not in direct communication in the way most of us experience. Visual information, for example, is not communicated from the right to the left or vice versa. The same with smell, touch, and sound. Since hemispheric control of the body is also largely contralateral (the right hemisphere controls the left side of the body, and the right brain the left of the body), it turns out that patients can only manipulate objects with their right hands if the commands come from the left hemisphere, and vice versa. (Either hemisphere can control the muscles of both upper arms, however.)

Over the years Gazzaniga has worked closely with a particular patient named Joe, whose corpus callosum was severed after he began having terrible seizures at age nineteen. Joe hasn't suffered any problems as a result of the surgery, and like most split-brain patients has gone on to live a normal and seizure-free life. However, because of the operation the two hemispheres of Joe's brain are not in instant communication. This isn't usually noticeable, except under specific conditions that have revealed some remarkable insights into how our minds work.

In one experiment with Joe that I witnessed several years ago, he was seated in front of a computer screen. Gazzaniga asked him to stare directly at a dot in the middle of the monitor. As Joe did, an image of a tree was flashed on the right side of the screen, and on the left the word *BLOW*. They appeared simultaneously, but for only a split second. Fixating on the dot in the middle ensured that Joe's left eye only saw the word *BLOW* and his right eye only registered the image of the tree. That meant that the image captured by Joe's right eye was sent to the verbal, left side of his brain, and the word *BLOW* was sent to the mute, right side.

After the experiment Joe was asked what he saw, and he immediately answered a tree because the left, speaking side of his brain had registered the image. But when he was asked what he saw on the left side of the screen, he said he had missed it. The fact is that he hadn't missed it. His eye registered the word and sent it to the right side of his brain, but it couldn't put what he saw into words because it had no capacity for language.

In another experiment Joe was asked to close his eyes. Gazzaniga then put a roll of tape in his left hand. He held it and rolled it around in his palm several times, but when he was asked what he thought it was, his best guess was a pencil—really nothing more than a stab in the dark. But when the same roll of tape was placed in his right hand, the left side of his brain immediately enabled him to identify it accurately.

These experiments mainly proved that Joe's two hemispheres weren't in direct and instantaneous communication. But the next experiment illustrated the powerful ways that the speech centers in the brain paint a picture of the world that each of us recognizes as reality, and how that picture may also play a crucial role in generating a sense of self.[22] Joe sat in front of the same computer monitor. This time the word *ORANGE* was flashed on the left side of the screen, where the signal traveled to the right, nonverbal hemisphere of the brain, and the image of a bird was flashed to the right. Next Joe was asked to draw a picture of what he saw with his left hand (the one controlled by the nonverbal side of the brain). Joe promptly picked up an orange marker and drew not the word *ORANGE* but a picture of an orange. When he had finished, however, he was perplexed. He really had no idea why he had drawn such a thing. When asked what he saw, he explained he saw a bird, not an orange (that was what registered in the left, speaking part of his brain). Gazzaniga then asked him to finish drawing what he saw with his right hand. Joe then converted the orange he had drawn into a bird that looked something like a kiwi.

Why was the bird orange? he was asked. Joe thought for a bit and then said he wasn't sure, just something he had come across, maybe a Baltimore oriole, he ventured. He literally couldn't say because he could not articulate the image that had been flashed to the right side of his brain. His reaction was reminiscent of Alice's remark in Lewis Carroll's *Through the Looking-Glass* when she says, "Somehow it seems to fill my head with ideas—only I don't exactly know what they are!" But he was trying mightily to say what they were, to come up with a reasonable explanation for the image he had just drawn, an action that to him frankly must have appeared completely crazy.

Based on experiments like this, Gazzaniga has developed a theory about how our minds work and the role the brain's speech centers play in it all. Joe taps the verbal side of his brain to explain behaviors set in motion by other nonverbal thinking he does but can't fully comprehend because the two hemispheres aren't directly connected by the corpus collosum. But since his behavior reveals that he is clearly having these experiences (the image of a bird flashed to the right hemisphere of his brain), how does he explain them? Basically he wings it.

"The left brain constructs a story to help explain the actual behaviors that are pouring out of the right brain," says Gazzaniga. "It ties it [the behavior] into a coherent whole."[23]

Why would Joe fabricate these elaborate stories to explain his behavior? Basically Gazzaniga's answer is: because he can, and he must, to make sense of his experience. And, says Gazzaniga, this is something all humans do.

His thinking follows along these lines. We have modules in the brain, clusters of neurons that have evolved over time to serve different purposes. They sense danger, react to fear, help solve problems, send and receive messages. These modules are like the additions built onto an old house—a dormer here, another bedroom, an expanded kitchen. All of these cerebral embellishments serve their purpose, and they all experience the world in their own ways, yet they are all also interconnected. They process information and experience feelings. In a sense they are multiple minds, each with its own peculiar slant on the world—aural, visual, emotional, intellectual, visceral. But they have one shortcoming: They cannot articulate what they are experiencing because they are not verbal—they evolved before language did. However, says Gazzaniga, the more recently evolved, vocal part of our brain *can* speak for them and their experiences, and for the actions and experiences of others. It may not speak accurately about them, but it can speak, and it does. He calls this part of the brain "the interpreter."

And so throughout the day we act and feel based on the experiences of other unconscious parts of our brain, and as a result we may grow elated or depressed or suspicious without any clear reason. The sources of those feelings could be as varied as cloudy skies, a favorite song, an old memory, an unconscious but crippling fear, even the body language of our boss or spouse. These ripple into our conscious mind, and the interpreter makes up stories to explain them to ourselves or to others. It translates them into verbal symbols and rationalizes our behavior. We need those rationalizations to make sense out of the world. Or as Jeff Goldblum's character in the movie *The Big Chill* put it, "Don't knock rationalization . . . I don't know anyone who can get through the day without two or three juicy rationalizations."

But just because the interpreter can tell us stories about our experiences doesn't mean the explanations resemble reality. After all, we don't know where most of our thoughts, actions, and feelings come from. But when they force themselves into our consciousness, the interpreter is driven to explain them. "The interpreter demands an explanation for felt states and behaviors. It won't let go," says Gazzaniga. "It is an absolutely crucial element [of the human animal], and there really isn't any evidence that any other species does it."

Perhaps this is why we appear to be the only self-aware animals: the speaking part of our mind bestows upon us that elusive thing we call our "self." It may be a grand illusion fabricated by clusters of neurons, but neurons that together constitute the single voice that tells each of us that we are of one mind, even though we are in fact of many. If it weren't for this voice we would all suffer from a kind of specieswide case of schizophrenia, or multiple personality disorder. Or we might experience life as a series of disconnected events with no "self" to experience, symbolize, or reflect on them. And bankrupt of language and speech, and the interpreter that they make possible, you and I would be without that voice in our heads that tells us, "This is you talking."

. . .

Whatever the case, when this voice arrived, so had the first members of our species, *Homo sapiens sapiens* some 195,000 years ago, according to theorists.[24] The first of our kind looked precisely as we do today. The sloped, ridged brows were gone. So were the snout-shaped jaws and hirsute bodies. Our legs were long and straight and our hips slim. All of the neurons that could be crammed into our skulls had been. If we had added any more, birth would have become impossible. But despite the limits that nature had placed on the size of our brains, our evolution didn't stop. It simply shifted playing fields and found new ways to store knowledge and information *outside* of our heads as well as inside. Carl Sagan once called this "extrasomatic memory." One also could call it human culture.

Strangely enough, human culture did not arrive at precisely the same moment our species did—at least not based on the fossil evidence we currently have in hand. The first glimmerings, in the forms of early sculpture, painting, and sophisticated tools, doesn't appear until about 50,000 years ago, when *Homo sapiens* began to migrate into the Middle East, Europe, Asia, and Australia.[25]

It is unclear why 145 millennia had to pass before we created our first works of art. Possibly the fossil record is incomplete. Maybe the magnificent cave paintings discovered in France, Spain, and Australia, and the weaponry and artwork that we have so far found are only part of a larger trove of artifacts that have so far eluded our discovery. Or perhaps time has simply obliterated them.

Whatever the case, even though we *looked* human 195,000 years ago, it appears it took some time before we began to *act* human. That could be because we hadn't yet mastered the spoken word. In fact, it's unlikely that without modern articulate language we could never have tethered our minds together to construct the colonies of ideas needed to develop the foundations of economies, trade, agriculture, art, religion, and science.

Derek Bickerton and Noam Chomsky, two giants in the field of linguistics, have subscribed to the "big bang" theory of language as a possible explanation.[26] It could be that 50,000 years ago the prefrontal cortex completed the neural pathways needed to gather all of the brain's different modules into a unified whole rather than a confederacy of clustered neurons that could only murmur chemically and unintentionally to one another. Maybe as these final links connected, they flicked a neural switch, reached a kind of interconnected critical mass, and the first truly human mind, complete with all its linguistic and artistic capabilities, turned on.

Or perhaps our pharynx was ready to operate and our brain was perfectly capable of conjuring and expressing symbols much earlier, but it took tens of thousands of years before we learned to control the hundred muscles in our throats, lungs, and mouths needed to utter the things we call words.

Possibly we simply needed time to build the language structures that could harness enough minds together for human cultures to begin to grow. The speech of children, immigrants, pidgin speakers, and inexperienced tourists prove every day that we can manage to communicate marginally even when we don't have total command of a language. Perhaps we verbally limped along, improving in bits and pieces just as a child does, until finally we mastered the process and accelerated the business of building cultures.

Merlin Donald has wondered if between 195,000 B.C. and 50,000 B.C., with our brains now full-size, we were still working out the kinks of human creativity and communication. He suspects that social interactions were central to the evolution of language, and wonders if the shift into a higher gear might have been driven by our underlying need to explain the workings of the world around us now that we were more aware of them. He calls this "mythic culture" and points to the "stone age societies" whose technologies have hardly changed in 35,000 years. They were discovered over the past century and a half: the natives of Tasmania, the Tasaday of the Philippines, the Bushmen of South Africa, and Pygmies of central Africa, each a culture that developed elaborate tribal rules, myths, rituals, and language. He points out that even though there has not been much technological progress, there has been very elaborate social progress. They had the minds to model and imagine mythic explanations of how the world works.

Donald calls all myth "the prototypal, fundamental, integrative mind tool." On close inspection Donald's integrative mind tool bears a striking resemblance to Michael Gazzaniga's "interpreter." The interpreter's stories not only get us through the day, but are also the stuff out of which we have erected the buttresses that support human culture. Our overwhelming need to explain our own behavior and to make sense of the mysterious events all around us is the source of our earliest myths, which are also rationalizations, except they are elaborate and culturewide, accepted by many. They are woven to explain how the world came to be, why we are here, where we go when we die, why the sun rises each day and the moon changes its shape as the weeks pass. Religion, literature, philosophy, and every branch of science owe their existence, one way or another, to the interpretive part of our brain that is obsessed with answering the question: Why?[27]

The importance of myth and interpretation, which are the direct results of our mind's ability to verbally explain our experience, also reveal that we are not purely intellectual creatures. Our fears, joys, and passions drive the stories we tell ourselves about the way we and our world work. We are human beings because we are also emotional beings.

The traditional view of human evolution has been that as we grew more intelligent, we increasingly left our primal drives in the ancestral dust, shaking off the shackles of emotion and rising to a better self. In fact, that notion is backward. Our increased intellect hasn't placed more distance between us and our old drives, it has amplified, reshaped, and enhanced them. Our emotional life is more complicated and enriched *because* of our intelligence, not because our intelligence has obliterated our less intellectual side. In fact, our big brains have *created* the immense emotional life we all enjoy. Primal drives that in a simpler being once largely focused on fight or flight, fear, hunger, satisfaction, and procreation have been transformed in us into complex emotions: love, hate, affection, friendship, jealousy, and every other possible combination of sin and virtue.

So evolution hasn't discarded the drives and behaviors that enabled our ancestors to survive and us to emerge. Instead it has preserved and built upon them. This is why we even *have* to rationalize the behavior we see, or fabricate myths that explain the world and our own behaviors. Those primal, nonverbal fears and feelings are what beg for explanation.

Yet there are parts of us that language, powerful as it is, can never fully touch or express. Having evolved before conscious expression, they elude words. And so we have also developed other, newer ways of communicating even after language evolved. These reach back through time to the primal parts of our nature, but link to the newest parts of our brain to send the most intensely human information we can share. Over time they have developed into three extraordinary traits—laughing, kissing, and crying. Each is a mysterious, and wordless, form of communication. Each belongs to us and us alone. And each is a testament to our profoundly human need to hold on to one another.

IV
Laughter

Chapter 8

Howls, Hoots, and Calls

Two cannibals are sitting beside a large fire after eating the best meal they have had in ages.

"Your wife sure makes a good roast," says the one cannibal.

"Yeah," replies the other, "I'm really going to miss her."
—*Anonymous*

L AUGHTER IS ONE OF THE GREAT MYSTERIES of human behavior. It evades understanding and resists analysis, partly because it thoroughly combines the primal and the intellectual parts of us. Yet we barely acknowledge what an unusual behavior laughing is, mostly because it is so woven into the woof and weave of our lives. Like the noses on our faces and the lobes of our ears, it's familiar to the point of invisibility. Yet if it were suddenly plucked out of our existence, we would be lost because we use it constantly to send strange and mysterious signals to one another.

The origins of laughter are ancient and wordless, a behavior whose roots run a good deal deeper than the evolutionary wellsprings of language. It is related to play and feeling good, although it isn't simply about fun. Darwin observed that it can also show up when we are feeling anger, shame, or nervousness, acting to mask, rather than display, emotion. At other times it may communicate appeasement or submission.[1] Or as Dante put it, "He is not always at ease who laughs."[2]

The social nature of laughter also makes it contagious. When someone laughs, the rest of us almost always do the same. This is why Charles Douglass, a television technician, invented the laugh track in 1953, and why it is still effectively used today to make some sitcom jokes seem funnier than they actually are.[3] It is why, even when we watch total strangers laughing about something we know nothing about, we will involuntarily smile or chuckle.

Maybe this explains laughter's universality. Everyone laughs, no matter where he lives, no matter what her race or background, whether he hunts corporate heads among the skyscrapers of Manhattan or real ones in the rain forests of Borneo. It sews us together as a species and as people. Along with big toes and thumbs and our oddly designed throats, it is one of those unique traits that distinguish us from the other animals.

Despite its familiarity and universality, we are almost entirely clueless about how we have become the laughing creature. There is no obvious, practical reason for laughing. If evolution resolutely favors the emergence of the eminently practical, what possible purpose could laughter serve? It's loud and calls attention to us—not necessarily a good thing when avoiding carnivores on the savanna, or hunting mammoths on the tundra. And when we laugh we tend to lose control, as though our minds and bodies have been hijacked, also not a recommended survival technique. Nor is laughing at predators really a very good idea, at least not unless it is done well after the hunt, by a campfire, in a cave far, far away.

In the crucible of evolution, behaviors also tend to become increasingly entwined in the traffic of other complex behaviors, and after a while it becomes maddeningly difficult to unravel one from the other. So we find it hard to know if laughs came before smiles or the other way around. Why one thing *sounds* funny while another *looks* funny? Or why laughter always surprises us? Understanding the origins of laughter requires a kind of psychological archaeology that forces us to compare glimpses of our closest primate relatives to the careful observation of ourselves.

. . .

A couple of years ago, a British team of scientists at the University of Hertfordshire headed by psychologist Dr. Richard Wiseman decided to find out what, exactly, people around the world found truly funny. They called this undertaking the LaughLab project, and set up a Web site where they invited people to submit their favorite jokes, while at the same time they asked them to rate other jokes that had already been posted. Within days LaughLab.com became one of the top ten Web sites in the world. Some days as many as 3 million people hit its pages. In the end 350,00 people submitted 40,000 jokes and 2 million ratings.

After an exhaustive and presumably hilarious analysis of all the information, LaughLab revealed to the world the joke that had been rated the funniest of all. Here is the choice:

"A couple of New Jersey hunters are out in the woods when one of them falls to the ground. He doesn't seem to be breathing; his eyes are rolled back in his head. The other guy whips out his cell phone and calls the emergency services. He gasps to the operator: 'My friend is dead! What can I do?' The operator, in a calm, soothing voice says, 'Just take it easy. I can help. First, let's make sure he's dead.' There is a silence, then a shot. The guy's voice comes back on the line. He says, 'Okay, now what?' "

Why was this joke chosen as the funniest? Because of its broad appeal, said Wiseman. Men and women liked it, young and old, Belgians, Germans, Americans, British, whoever read it. Wiseman theorized that one appeal was that readers felt superior to the less than brilliant hunter who made the phone call. Another was that it provided a release about people's own fear of death, something we all share.

As far back as 1905, Freud made a similar observation about jokes. He said they masked or released, in a socially acceptable way, fears and feelings that might be otherwise inappropriate. And a laugh, he speculated, physically revealed the relief felt when something disturbing was expressed. This, he said, was similar to what we do in our dreaming—unconsciously conveying what we are not entirely comfortable with consciously.[4] In other words, there was a link between laughter and the subconscious. Maybe jokes and one-liners functioned like dreams because they often originated in the obscene or aggressive tendencies we saw in our mind's eye as we sleep. Something dark and startling, maybe even angry, lurked behind every joke. But the joke itself made the expression of that darkness acceptable because it disguised it as positive. And the final result was that elusive feeling we call funny.[5]

LaughLab didn't concur precisely with Freud. It concluded that the most compelling reason why so many people found the hunting joke funny was its sharp incongruity. After all, you figure that someone concerned enough about his friend to call 911 in the first place would go back and delicately check his pulse, not shoot him. That's completely unexpected, and under the circumstances, hilarious because the whiplash is so sharp.

Darkness in humor and whiplashes in logic seem to be the sources of the most central aspect of laughter: surprise. To be amused, our minds have to be headed in one direction and then yanked unceremoniously in the other.[6] In that moment of surprise and confusion, as our neurons struggle to process all of the conflicting cues and resolve all of the mixed messages, we suddenly "get" the joke. This is why you can never precisely predict a laugh because we never consciously decide to do it in the first place. Never. A premeditated laugh is a fake laugh, and most of us recognize it instantly when we hear it.

Wiseman's LaughLab didn't simply observe the whiplashing effect jokes have on us, it actually measured it, using functional magnetic resonance imaging to peer into the brains of people as they listened, first, to the initial part of various jokes, and then to the punch lines. Later they compared those images to scans recorded when the same subjects were simply listening to a series of unfunny sentences.

The scans found that humor and laughter are scattered throughout the brain. There is—cerebrally speaking, at least—no comedy central. We have clusters of neurons devoted exclusively to all sorts of different aspects of laughter and humor—areas for hearing, seeing, and recognizing the laugh of someone else, sectors that distinguish between puns and slapstick humor, and special neurons that send signals to our lungs and pharynx so that we don't simply *feel* amused but actually laugh. Yet none of these are bunched together in one spot. This indicates that human laughter has evolved over time, changing from something old to something new, connecting more recently arrived parts of the brain with older areas.

For example, a semantic joke such as. "What don't sharks bite lawyers? Professional courtesy," is initially processed in the temporal lobes, which sit roughly above our ears. Puns, on the other hand, like "Why did the golfer wear two pairs of pants? Because he had a hole in one!" are first processed near Wernicke's area on the left, verbal side of the brain, presumably because puns are verbal by nature.[7] These are the different areas of the brain where we sort through all of the raw information we register and then work out the basic meaning of what we are hearing. But that is only part of the process.

If people find a joke funny, LaughLab's MRIs revealed that a very precise area of the brain just above the right eyebrow called the ventromedial prefrontal cortex suddenly lit up. This cluster of neurons is as close as the brain gets to a funny bone. Several brain-scanning experiments have shown that this is where we "see" incongruity, and then register the surprise that makes us laugh.[8] It's the part of the brain that "gets" the joke.[9]

It is not, however, the part of the brain that experiences the sensation we call funny. That is situated in still another place, far away, near the base of the brain in an area called the nucleus accumbens, a location, logically enough, associated with positive emotions in animals and identified as a key site in moderating drug addiction. Its location is so close to that area, in fact, that some researchers have wondered if it might help explain why we can never get too much of a good laugh. Fun has its addictive qualities, too, after all.

There is a final sector of the brain that actually triggers laughter. And it resides in still another location—the selfsame area, in fact, that helps us direct our thumbs and fingers to make tools, and our lungs, throats, and tongues to make words. Scientists call it the supplemental motor area, or SMA. Located near the top of the brain, it was first pinpointed in the late 1990s, when researchers at the University of Rochester School of Medicine performed four humor tests on thirteen people at the same time they scanned their brains. In the first test, each subject was asked to laugh along with the people they heard laughing in a recording. No jokes were involved. It was like listening to a laugh track without a sitcom. In the next experiment the patients were asked to listen to the laughter but not to laugh along. In the third test, they each read written jokes, and in the fourth they viewed a series of wordless cartoons.

Whenever anybody laughed for any reason, the MRI machines indicated that the SMA was always activated. It seems that this area plays a central role in all kinds of movement—hands, feet, legs, even eyes. In the case of laughter, it gathers up those disparate signals from other parts of the brain that say it is time to laugh and sends impulses to our throats and chests and the fifteen muscles our faces use when we laugh.

The odd thing is that this part of the brain can trigger laughter even when something isn't really funny. Surgeons found this out at the UCLA

School of Medicine a few years ago when they were performing exploratory surgery on the brain of a sixteen-year-old girl who was suffering from intractable seizures. They placed electrodes in different parts of the girl's brain to locate the sections that were causing her problems. When neurons were stimulated near the supplemental motor area, they found that she often laughed, even though there wasn't anything particularly funny going on. When the girl was asked what she found so funny, her answers were reminiscent of the split-brain patients of Michael Gazzaniga who made stories up to explain why they were drawing odd pictures (see chapter 7). If she was reading a story when the brain stimulation made her laugh, for example, she said the passage she was reading was funny, even if it wasn't. One time when she was simply showing the doctors she could touch the tips of two fingers together, an electrical stimulation made her laugh. When asked why, she answered, "you guys are just so funny . . . standing around."[10] This might also help explain why some studies have shown that even when we don't feel happy, smiling can make us feel happier. Our minds apparently tell us, whenever we are laughing or smiling, that we must be happy, and so then we are.

There is also the strange case of a landscaper from Iowa known in the scientific literature as CB. His affliction recalls those juxtaposed incongruities that seem so important to anything we find funny. CB suffered a stroke at the unusually young age of forty-eight. Thankfully he made a full recovery, except for one perplexing problem. He occasionally, and for no apparent reason, would break into uncontrollable fits of laughter. When this happened, he was almost never experiencing anything funny. In fact, he wasn't even thinking an amusing thought, nor was he in the company of anyone else who was so much as chuckling. Yet great overwhelming storms of laughter would ambush him. Not only that, at other times he suddenly found himself weeping uncontrollably. Again, he didn't feel any of the emotions that normally cause us to cry. It just came on, like a cloudburst.

The affliction that CB suffers from is common enough that it has its own medical acronym, PLC—pathological laughter and crying. All victims of PLC have, for one reason or another, had very tiny parts of their brains damaged related to the supplemental motor area and the pathways through which it dispatches its signals. Somehow or other, neurons are firing that say laugh or cry, but they aren't being prompted by the parts of the brain that actually process the emotions that cause genuine laughter or tears.

Most of us don't suffer from this problem, but it's interesting that a central theory about the origin of laughter and the whiplashing effect that is perceived in our prefrontal cortex can be traced back to the neurological locations associated with laughing and crying. On closer inspection, it turns out that there might be a good deal more to this connection than coincidence.

. . .

Darwin noticed that the facial expression of a person laughing often looked identical to someone who was crying. But it was the British zoologist Desmond Morris who first speculated in his popular book *The Naked Ape* that the origin of laughter might actually be directly linked to crying.

In the earliest months of life we have one primary way of expressing fear, loneliness, pain, or any other discomfort: We bawl, long and loudly. It is a simple way of making our point, but highly effective. Any parent can tell you that.

During our first ninety days, we do this indiscriminately. Pretty much any adult will do as long as he or she fixes the baby's problems—fresh diaper, food, warmth. At this point in life, all faces appear neutral, and therefore good because infants at this age simply don't have the cognitive firepower to know the difference between a familiar face and a strange one.

But then around the fourth month, at some very basic level, the brain begins to wire up connections that enable us to recognize the primary caregivers in our lives. This is also the age at which we begin to smile and giggle, an immense milestone in human relationships. Mothers and fathers live for a child's first laugh, because it marks their offspring's relationship to them as personal, not just practical; as if the child is saying, "I know you! You are special!" Their child is reacting not simply to basic, faceless needs like an empty stomach, but to Mom and Dad themselves. That produces a powerful bond.

Looked at this way, you can view a baby's laugh as a survival technique. Since human babies are essentially born twelve months early compared with other primates, we enter the world the most helpless mammals on the planet. We require prodigious doses of parental commitment and care. We are easily injured; we are unable, literally, to hold our heads up for months and incapable of navigating. We demand coddling, feeding, and attention day and night.

But a baby's laugh provides a powerful emotional gift that encourages her care. The more laughter, the more bonding; the more bonding; the better the chances of survival. It sparks a potent feedback loop. This may explain why across every culture, parents play with their babies to encourage them to laugh.

But what could have put this play-laugh feedback loop in motion in the first place? Morris explains it this way: imagine you are a prehistoric four-month-old cruising along in the arms of your mother when suddenly you are startled. Being four months old, your immediate reaction is to cry. But in another split second you realize everything must be okay because you are firmly and safely in the grip of your mother, and your mother reassures you with a coo that everything is fine. Suddenly your fears quickly reverse, and you are utterly relieved. You interrupt your crying with a yip of recognition that chops the incipient long, loud bawl down into smaller ha-ha-has.

As Morris envisioned it, this combination of mixed input sent a seemingly contradictory message: (1) There's danger, but (2) not really; you are fine. This is the whiplash effect at its most basic. Maybe, he has speculated, laughter evolved from this combination of alarm and relief.[11]

The thing about play is that it, too, is from an evolutionary standpoint, all about incongruous opposites. It seems to be about fun, but really it's a survival technique, like laughter itself. Mock fighting, biting, tumbling, and chasing among young mammals provide a kind of dry run for the real battles that will follow.

A big part of playing among primates involves tickling. Young chimps and gorillas spend a lot of time getting tickled—by their immediate family, or by one another. This has caused some researchers to propose that the building blocks of humor might be built upon the foundations of nature's tickle reflex.[12] Tickling, after all, exhibits many of the same dynamics that more sophisticated humor does. There is the element of surprise, for example, which may help explain why it's impossible for us to tickle ourselves (your supplemental motor area won't allow it). Tickling also juxtaposes danger and safety, pleasure and discomfort. It is a kind of mock attack. Neurological tests even indicate that the sensation of a tickle travels simultaneously along two separate sets of nerves fibers evolved to register completely opposite sensations: one for pleasure and one for pain. And finally, tickling is interpersonal, as all humor is. It takes at least two to tickle, just as it does (usually) to laugh.[13] Or does it?

Scientist Christine Harris was so fascinated by the question of the personal nature of tickling that she set up an unusual experiment designed to figure out if laughter, even ticklish laughter, requires two people. Her hope was that by figuring out the nature of ticklish laughter she might also get a handle on how every other kind of laughter works.

But how do you test whether human contact is necessary to make a tickle ticklish when tickling is always done by humans? In other words, how do you take the human out of tickling? Harris and her team decided that creating a machine made for tickling might do the trick. If ticklish laughter had to involve at least one other real human, she figured, then no one would laugh if a machine tickled them.

That, however, raised another problem: How do you create a machine that can tickle, and how do you make sure the person being tickled doesn't know whether it is a machine or another human being doing the tickling? In the end, the researchers faked it. They created a mock tickling machine in their laboratory, complete with a robotic-looking hand, a vacuum-cleaner hose, and a nebulizer used in asthmatic therapy to provide some convincing sound effects. But the hand wasn't actually robotic at all. It couldn't even move.

Subjects were told they would be tickled twice, once by a human experimenter and once by the machine. Next they were blindfolded, ostensibly to help them better concentrate on the tickling sensation. All of the tickling was actually done by a human hand that belonged to a second experimenter who was hiding beneath a cloth-draped table right next to the subject. The subject was told that he or she would be tickled once by the machine and once by a human, and the experimenter was careful to tickle the same way. If the subjects laughed when they thought they were being tickled by a human, but not when they thought they were being tickled by a machine, that would indicate tickling required human contact. It turns out it didn't matter. Whether they thought it was a machine or a human tickle, the subjects laughed just as much. This was true even when the subjects were left in the room alone with the "machine," and thought there were no humans anywhere nearby.

But maybe the body somehow knows a human tickle when it feels it no matter what, and each of the subjects unconsciously recognized that the machine tickle was actually human. There's no way to know. Whatever the case, it does seem that ticklish laughter and humorous laughter are cousins, even if we might not think so. Ticklish laughter, after all, doesn't seem to involve any sort of wit or humor, but it does involve mixed messages and the same whiplashing opposites. Maybe in tickling we can glimpse a simpler, more primal form of humor, a foundation upon which the higher, cognitive experiences that make us laugh are built when we hear or see something funny. Maybe visual and verbal humor are a kind of symbolic tickling, except that rather than tickling our feet or stomach or neck, they tickle our minds.

. . .

There are other ways that tickling and play, and play and humor may be linked. Even though we resist tickling, we also only allow ourselves to be tickled by people we trust. Darwin pointed out that any child tickled by a total stranger would be unlikely to laugh. In fact, he predicted she would scream in terror. Tickling is intimate; it allows us to bond more closely with people we are already close to. The same happens when we laugh and joke with friends or even with strangers; we are connecting with them. This makes laughter a powerful form of nonverbal communication that inevitably draws us together. Considering how much we need one another as a species, there is enormous value in that.

But how did laughter, and the feelings and emotions we experience when we laugh, come about? Why do we make the particular sounds we do when we laugh, and why do we contort our faces? And finally, why does laughter always surprise us? Put another way, how did laughter evolve?

When chimps play, when they tumble and chase and tickle one another, they make a very precise panting sound, like rapid breathing. Some scientists call this laughter, but it is really nothing like the human variety. This doesn't mean, however, that they aren't connected. Robert Provine, a psychologist at the University of Maryland who has closely studied the origins of laughter, believes this panting evolved out of the way chimps breathe when they are worn out from play. He suspects that their efforts to catch their breath eventually developed into a kind of ritualized communication that also said "I'm having fun. I'm playing with you, not fighting." If this is the case, then physical reaction to tickling may eventually have evolved into a symbolic reaction that we recognize today as human laughter.

Still, the pants of chimps are not the same as a human laugh. Chimp pants, like all panting, is a sound made when air is inhaled *and* when it is exhaled. But human laughter is only exhaled. One explanation for this is that we have far more precise control of our vocal tracts when pushing air than when we pull it in. (Try talking as you inhale.) As a result, when we talk we inhale rapidly to set up the longer phrases we form when we exhale. This basically means that we have subordinated our breathing to our speaking. It may also explain why we don't pant like tickled chimpanzees when someone gets off a witty remark. All of our laughs are exhaled (though we sometimes gasp for breath when we are laughing really hard).

Provine theorizes that the reason we don't pant like chimps is because we walk on two legs, not on all fours. He came to this conclusion after recording and combing through the sound structure of countless human laughs. He found that no matter whether we titter or guffaw, we generally chop laughs into outward blasts of air that last about a fifteenth of a second and then repeat those about every fifth of a second.[14] But regardless of how often we do this, the sound is always exhaled. This is what gives human laughter its unique ha-ha-ha, staccato rhythm.

Chimpanzees, however, don't chop their pants into short little pieces. They are longer and deliver one pant per breath. They do this because, like most four-legged creatures (remember, they mostly knuckle-walk), the pattern of *their* breathing is constrained by the rhythm of their walking. One step, one breath.[15] Chimps don't have the same precise control over the lungs and the muscles that control their breathing that we do. Their breathing and play-panting are related to their four-leggedness.

Provine says we have been able to commandeer these breathing muscles because upright walking released us from the one-step, one-breath formula.[16] If he is right, this means that speech has left its signature on the sounds and rhythms of our laughter; even though they have very separate origins. We sound the way we do when we laugh, partly because we evolved the ability to speak.[17] And not only that, we speak and laugh the way we do because our big toes enabled us to stand upright and learn to breathe differently in the first place.

· · ·

The facial expressions that go along with our laughter are another matter, although these also trace their origins back to play and primal forms of communication. Not that the path to these origins is even slightly straight-forward.

When chimps are truly threatened or angry and ready to attack, they pull back their lips and completely bare their teeth. They scream and hoot and generally make as much of a ruckus as they can. There is no real fore-thought that goes into these snarls, or any of the other antics that accompany them. It is all instinctual and unplanned, genetically driven.

Jan van Hooff, an ethiologist at the University of Utrecht, likes to make these points when he lectures while showing the images of a documentary film on human and primate behavior produced by Joost de Haas. He stands by the screen that juxtaposes images of chimps battling and cavorting, with film and pictures of humans laughing uproariously. At first glance these images appear to have nothing to do with one another, but in fact they are connected in some very surprising ways.

For example, when chimps are playing, they tone down the snarling, aggressive way they bare their teeth in an actual attack by dropping their lower lip over their fangs. This, Hooff says, is a signal that they aren't really angry. When they do this, their mouths also tend to look more like the way ours do when we laugh, which don't reveal as many teeth as when we snarl or scream at someone.

Other parts of the face also are involved. Chimps' eyes grow wider when they play, which is to say they aren't battle-intense. And their brow is unfurrowed. Both combine with the panting sound to reinforce the message that this is a mock battle, not the real thing. If somehow the characteristic sound of human laughter could replace the sounds of the chimp's panting, the resemblance between a laughing human and a playing chimp might look pretty similar.[18] However, they wouldn't be exactly the same for one simple reason: We humans, because we have so many more facial muscles than our cousin chimps and gorillas, can be much more expressive and far more subtle in our expressions. Nevertheless, it is possible to see how our "laugh face" evolved from the same expressions that apes use when they play. They, and we, modify the looks on our faces to say "just kidding."

The research that Provine and Hooff have done helps explain why we look and sound the way we do when we laugh, but it still doesn't explain why laughter always surprises us. As it turns out, however, there is a theory for that, too.

. . .

Try for a moment to listen to a laugh as if you had never heard it before. Out of context, it sounds animal-like—a wild call in the jungle or a secret, primal message sent between two members of an inhuman species. There's a reason for that. Our laughter has far more in common with the screams of an excited chimp than the eloquence of Winston Churchill. That trait makes it both simultaneously simpler and more mysterious than words. With speech, we form thoughts and then more or less purposefully shape words to express those thoughts. But laughter happens the other way around: We laugh as the result of being mentally and emotionally ambushed.[19]

This also happens with other primates and their calls, though the calls aren't related to humor or even play. When chimps forage in a forest, for example, and come across food, they involuntarily call out in a distinctive way to let the others nearby, usually siblings or immediate family, know they have found something good to eat. This is a little evolutionary trick, a way of upping the survival rate among the family members of a foraging group and spreading the wealth. It is also a kind of sound symbol that says "Over here, food!" But unlike language, it is not learned; the calls are inborn, acquired genetically, and they just erupt in the way most dogs bark when they sense danger. It can't be helped.

Jane Goodall tells a great story that reveals just how uncontrollable these sorts of calls are.[20] While doing her research, she hid a cache of bananas for the Gombe chimps she lived among near an observation area. One day a particular chimp came upon this fabulous windfall and began to make the involuntary food call. But as the call was escaping its throat, the chimp cupped its hands over its mouth and tried to stifle the sound, the way you and I might try to stifle laughing at something funny we see in church or at a funeral. But this didn't stop the call any more than we can sometimes stop laughing even when it's inappropriate. So he stood in the forest, involuntarily informing every chimp within earshot of his tremendous good fortune even though he had slapped his hand over his mouth. Aeons of evolution had the upper hand, so to speak, and the word got out.

Laughter is the same. We never see it coming, and when it does, we have limited ability to stop it. So it seems that when we laugh, we have one foot in the primal world and one in the modern human one, the one that requires sophisticated intelligence.

Recent studies have shown that laughter can be a powerful healer. When we laugh, the brain and endocrine system release cocktails of pain-killing, euphoria-producing endorphins and enkephalins as well as dopamine, noradrenaline, and adrenaline. All of these not only put a smile on our faces, they actually make us healthier because they contribute to a strong immune system.

Endorphins are naturally occurring neurochemicals that kill pain and keep all sorts of discomforts at bay. Some scientists, for example, theorize that people who suffer from severe headaches have lowered levels of endorphins in their systems. Endorphins manage to mitigate pain because the amino acids in them attach to receptors in the brain and spinal cord. When they do, they block impulses that send pain messages from various parts of the body to the cerebral cortex. These are precisely the same receptors in the nervous system that respond to morphine.

Enkephalins have a similar effect. Like endorphins, they block pain. Opium has such a powerful effect on us because its chemical structure is remarkably similar to many enkephalins. Dopamine, on the other hand, is not a painkiller, but your brain can't function properly without it. Those who suffer from Parkinson's disease are victims of low dopamine levels. Obviously, having sufficient supplies on hand is a good thing. Finally, noradrenaline is a neurotransmitter that calms us and reduces stress. When enough of it is in the brain, it keeps the mind from going into overdrives of worry and anxiety.

The black sheep in this family of chemicals that laughter releases is adrenaline, which is not calming, although surges of it temporarily reduce our sense of pain. The classic times it kicks in are when we are in flight-or-fight mode—when we are in a confrontation with the boss, or a drunk at a bar, or facing a hungry lion on the savanna. It is released by the adrenal gland and increases heart rate, relaxes bronchial and intestinal muscles, stimulates the heart, sharpens the mind, and generally prepares the body for action. It seems an odd chemical to release when we are happy, but it may explain why our metabolism kicks into high gear when we are laughing hard. And perhaps it is the part of laughter that represents the threat before we are relieved whether it is peekaboo or a razor-sharp punch line. Maybe it represents the neurochemically dark side of laughter.

The positive effects of the neurotransmitters that laughter releases are so powerful that a field of medicine has grown up around it. In the 1980s writer Norman Cousins came down with ankylosing spondylitis, a degenerative connective tissue disease so debilitating he could barely raise his fingers. Doctors gave him a one-in-five-hundred chance of complete recovery. Utterly incapacitated, Cousins was reduced to lying in bed watching television. And that, according to his account, was how he found relief. He noticed that his pain subsided and he had an easier time falling asleep after watching Marx Brothers comedies and *Candid Camera* episodes. At first he did it simply for a reprieve, but eventually he applied humor and laughter as a kind of treatment. And over time he found that laughter actually began to heal him. Eventually he did make a complete recovery.[21]

Cousins' story inspired some scientists to take a harder look at the healing powers of laughter. Today it turns out to be building an impressive track record. Research at the UCLA Medical Center indicates that not only can watching funny videos help children in the hospital better withstand painful medical procedures for cancer and other diseases and injuries, but it actually releases natural killer cells (NKs), a type of lymphocyte that routinely patrols the body for any cells that are infected or aberrant. Like police, they constantly look for trouble, and when they find it, they try to rub it out.

Obviously if someone is suffering from cancer or otherwise ill, the body will be subjected to more aberrant or infected cells, so anything that increases the effectiveness of NK lymphocytes is good. And this is apparently exactly what happens when we laugh. A study of the young patients at UCLA revealed that laughter not only generated more NK cells, it also made them more active and efficient. Scientists also saw an increase in disease-fighting B cells, immunoglobulin antibodies that fight respiratory infections, and an increase in a substance called Complement 3, which helps antibodies pierce and destroy the dysfunctional cells they attack.[22]

In another study, in Osaka, Japan, twenty-one young males were asked to watch a tourist video and then performances featuring Japan's most popular comedians. Unlike the UCLA study, the team didn't find an increase in NK cells, but they did find a 27 to 29 percent increase in their activity. In other words, NK lymphocytes didn't increase in number, but they were patrolling and killing aberrant cells with more vim and vigor than they were before the laughter. Either way, it seems that the writers of Proverbs had it right when they concluded four thousand years ago, "A merry heart does good like a medicine" (Proverbs 17:22).*

You might think we do most of our laughing when we are sitting in front of the TV or in a movie theater passively watching a comedian or some comic scene. But it's not true. The vast majority of the laughing we do is when we are simply enjoying one another's company. In fact, we are thirty times more likely to laugh when we are socializing than when we are alone.[23] This is because the purpose of laughter is all about social bonding and communication.

Another aspect of Robert Provine's work illustrates this. For more than a decade he and his students have eavesdropped on groups of humans hanging out, talking, and, above all, laughing in malls, bars, and coffee shops. Notepads in hand, they marked whether speakers were male or female,

* Another interesting finding by the scientists in Osaka was that the intensity and loudness of the laughter didn't affect the increase in NK cell activity. The better indicator was how good the subjects said they felt afterward. It was their positive mental amd emotional state of mind that correlated to the increase in killer cell patrolling. In other words, you couldn't tickle someone into good health. It was the positive feelings the laughter represented that did the job, not just the laugh itself.

who laughed when, who laughed most, and what was said right before the laughing began. His student teams came across some fascinating human behaviors, especially when it came to men and women.

Strangely enough, for example, Provine's research has shown that we laugh more when we are doing the talking than when we are listening—46 percent more often. And that when we are in mixed sexual company, no matter whether we are speakers or listeners, women laugh 127 percent more often than men. But men, when they are talking, laugh about 7 percent less than their female audience does. In other words, when human beings get together and laugh in mixed company, women do the vast majority of the laughing. And when females are talking, both the men and the women who are listening to them laugh less than they would if a man were talking.

It also turns out that laughter isn't so much about sidesplitting wit as it is about subtly and unconsciously lubricating the social process. Provine didn't find people clustered together getting off great Woody Allen–style one-liners or adroitly exchanging witty remarks. In fact, Provine found that only 20 percent of laughter is the result of a good joke or a killer one-liner. Mostly it was just a nonverbal response to friendly conversation—phrases like, "Look, it's André!" "Are you sure?" and "It was nice meeting you, too." Context and relationships and delivery were as important to the laughter as the actual things said.

In fact, in all of the recording and note-taking Provine and his battalions of students did, the funniest laugh lines they came across were, "You don't have to drink, just buy us drinks," and "Do you date within your species?" Funny, but not exactly the stuff of Groucho Marx or Noël Coward. But given the context, given the relationships, given the dynamics, and very likely the facial expressions, people laughed and enjoyed one another, and that was what was most impotant. They were connecting and cohering.

Laughing, says Provine, is about "mutual playfulness, in-group feeling and positive emotional tone." On this level you might say it sends a crucial message among the gathered humans that they are all, in effect, on the same page, and therefore in good and safe company. Imagine a group of people at a bar, congregating after work with one standing amid them impassive and unlaughing. Not scowling or aggressively contrary, merely impassive. This person would stand out like a roach on a newly frosted wedding cake. Nor would it be long before someone either made an effort to bring that person into the fold, or ask what was wrong. It is strange, even deeply disturbing, not to laugh when in the company of other laughing people.

The variety of ways we use laughter to communicate turn out to be far more refined than we might have thought, too. Sometimes we laugh because we don't know what else to do and it simply seems a safe enough response at the time, a mask more than a communication. This is probably the origin of the nervous laugh. Sometimes we laugh out of deference. It is a subtle way of giving the boss the spotlight or saying, "You're in charge." Or sometimes it is a way to mask an angry remark: "You look so much better tonight than usual."

That women laugh more than men also reveals another subtle aspect of laughter and the messages it sends. It probably doesn't mean that men are more innately witty than women. Instead it says something about the social evolution of our species. Remember Dunbar's theory that in mixed company, men talk more than women because they are lekking, the human equivalent of a peacock strutting. This might be another case of men performing for women, spreading their tail feathers and gauging the level of attention they are getting based on the laughter they generate.

A 1999 *Esquire* magazine poll revealed that more than anything, women looked for men who made them laugh.[24] On the surface it might seem to mean nothing more than a funny guy is a good time, which is true enough. But it also may mean, in the long run, that the laughter two people share is an excellent indicator of their compatibility. After all, to laugh together about something, you, and those laughing with you, must share the same values and see life through the same lens. Or to go back to the basic tenets of Theory of Mind, you are all empathizing with one another at that moment, imagining that the people around you are experiencing something very similar to what you are experiencing. You are of one mind. And that is comforting.

Two psychologists, Michael J. Owren and Jo-Anne Bachorowski, have theorized that laughter draws us together in strong and subtle ways the primal calls of apes do, not unlike the food-finding hoots that Jane Goodall noticed. They believe that like primate calls, human laughter evolved basically to get the attention of the others around you (something like the way a baby's crying gets the attention of a parent). Other forms of communication, like words, body language, and facial expressions, only serve to supercharge the effect laughing can have to make us more sexually attractive, dominant, friendly, accessible, wanted, or admired. If this is true, laughter may be more about influencing the behavior or perceptions of the people we are with than it is about humor. It is a way of saying, "I'm here, and you want to pay close attention to me. And if you are laughing with me, then I must be succeeding."[25]

This only adds another social facet to the mystery of laughter. If we turned the tables and imagined that we were creatures who never evolved to laugh, we might all have ended up autistics, largely incapable of calibrating our state of mind with the others around us. Autistic children have a hard time laughing appropriately because laughter requires that people engage emotionally. That's because laughter and the ability to put ourselves in another person's shoes go hand-in-hand.

Put another way, we bond with those we laugh with, and we laugh most with those we feel most comfortable around. We share our biggest belly laughs with our closest friends. Laughing together says that we are aligned, all part of the same clan, members of the troop. When we laugh together it is among the few times that we don't have to apply heaps of mental energy to imagining what those around us are feeling and thinking. For the moment at least it is not necessary to figure out what is on everyone else's mind because we already know; we find the same thing funny. It is almost as if, when we are gathered in social clusters, laughter has become a way to check in on those around us to see if they are thinking what we are thinking. It is a nudge that says, "You get that, right? I can trust you."

The effect is cumulative, too. The more we laugh in the company of particular persons, the more we trust them and the more they trust us. Whether you are twelve years old or ninety, laughter draws us together in ways that words can't begin to. We can tell someone we like that person a million times, but those words don't cement relationships with anything like the power of long evenings genuinely laughing with one another. As emotional glues go, it is extraordinarily strong. And that, when seen through the long lens of evolution, is an enormous relief.

V
Tears

Chapter 9

The Creature That Weeps

Why the heck do people cry? It is such a weird thing to do. You get upset and water comes out of your eyes.
—James Gross, clinical psychologist, Stanford University

The advantage of the emotions is that they lead us astray.
—Oscar Wilde

SCIENTISTS ARE COMPELLED TO ADMIT they really don't understand why we cry. They can only agree that we are the only animal that does. Other animals may whimper, moan, and howl, but none cries tears of emotion, not even our closest primate relatives. And unlike laughing or even speech, there doesn't seem to be any obvious analog in the primate world. Apes do have tear ducts, and so do other mammals, but they are there purely for housecleaning: Tears bathe and heal the eyes. We have similar plumbing in place to keep our eyes clean and disease-free, but for some reason, at some point in our evolution, a savanna ape, or perhaps an early version of our own species, developed a direct, physical connection between the gland that makes our tears and the emotional parts of our brains. That is unique in nature.

Like every genetic mutation, this connection was a mistake. But a mistake that worked, an adaptation that somehow enhanced the survival of the creature that ended up with the wayward gene, which it then passed along to other generations until it became not an aberration, but a successful trait.

When emotions overwhelm us, a small organ called the lacrimal gland, which sits just beyond the outer corner of each eye, can generate so many tears that they fill, and overflow the ducts that rim the southern hemisphere of our eyes. This is what makes human crying so unusual. Many animals can feel longing, fear, or pain. But it is the tears, combined with our emotions, that makes human crying unlike any other natural behavior.

The very first thing a baby does when he enters this world is cry. It is a primal, unmistakable announcement that he has arrived.[1] The birth cry signals two simple things: the child is alive, and the umbilical cord can be safely cut, which marks us as a whole and separate human being. During the first three to four months of life, before we learn to smile or laugh, crying is the primary communication method we have. But between eight and twelve months we begin to cry less as we develop other ways to express what we want—by pointing or grunting, or tossing spoons, cereal, and bottles around. But in our earliest days, we use crying often and with great effect.

An infant's cries work so well partly because parents' ears are attuned to the bawling of their children. Nature has rigged it this way. Human mothers can almost always distinguish the cries of their own infants from the cries of others. Babies even have different cries that send different messages—shrieks and screams of pain that mean something is seriously wrong, or cries of separation, discomfort, and hunger. Each cry becomes a kind of rudimentary, wailing vocabulary that precedes a baby's first words. In fact, some linguists have theorized that the rhythmic rising and falling pitch of an infant's cries form the basic intonation pattern for all human sentences, which normally begin on an ascending note and then end on a descending one. When these cries are combined with the pained, reddening, puckered faces that often go with them, the two can garner a lot of very focused attention. (This face, it turns out, resembles the "frustration-sadness," "whimper," and "cry" faces of apes.[2])

As we grow older our reasons for crying become invested with more subtle shades of feeling. Our pain and discomfort is no longer simply physical, but also emotional. And it is almost always inexplicable. We cry, it seems, because our emotions have outflanked the simple syntax of speech. Nouns and verbs and adjectives, and the logic that goes with them, simply aren't up to the job of explaining our feelings. If we could use words, we might not need to cry at all. But, of course, we do because crying, like laughter, is a primal form of communication that taps the emotional, mute, and unconscious parts of our brains and experience.

In fact, electromyographic studies show that the nerves that operate the muscle that makes our chin quiver when we are on the verge of tears (the mentalis muscle), or that put the lump in our throats or depress the corners of our lips (depressor anguli oris muscle), are all very difficult to consciously and purposefully control. Yet the slightest disappointment instantly reveals itself in the expression of downturned lips. In fact our mentalis muscle never really stops moving, which is another way of saying it is an entirely unconscious physical representation of our emotions. These nerves and muscles simply don't check in with the verbal and conscious parts of our mind before doing what they do. This is why even when babies are born without structures above the midbrain they can still cry, an indication that the feelings associated with crying run deep into our evolutionary history, long before the apparatus of speech and conscious thought emerged.[3]

. . .

Tears also serve a physiological purpose as part of the support system that operates our eyes. The rods and cones, the optic nerve, and the intricate geometries that enable us to perceive light are all amazing evolutionary innovations, but they couldn't function if it weren't for tears.

To look at them (and through them) the lenses of the human eye appear perfectly smooth, but in fact they are pocked and rutted and wrinkled, moonlike in their geography. But our tears fill in and smooth out the imperfections of our lenses every time we blink, an average of twelve times per minute. Without their steady drip, the world would look as if we were peering through a plastic bag, and we would have, roughly, the visual acuity of Mr. Magoo.

Tears do not consist entirely of water; they are actually a kind of "fluid sandwich" made of three separate layers. The inner layer, the part of the sandwich that bathes the cornea, or lens of the eye, consists of a lubricant called mucin. The middle layer is mostly water, and the outer layer, over which the lids of our eyes fold, is made of oils evolved to keep our tears from evaporating. If our tears were not incessantly lubricating and cleansing our eyes, we would quickly and painfully lose them to infection and disease.

Just as our tears have three layers, tears themselves come in three varieties: reflex, basal, and psychic. Each is unique in its purpose and its chemistry. Reflex tears that form when we get shampoo or a grain of blown sand in our eye, are produced automatically in the main lacrimal gland to flush the eye and help heal any damage. Basal tears stream continuously over our eyes so we can see clearly, and they dampen and remove dust and debris. Psychic tears are the ones that confound scientists. They well up when we experience strong emotion—mostly sadness, it seems—but also when we feel intense pride, anger, frustration, or love and warmth.

Whatever the reasons for our tears, they all flow from the same intricate lachrymal system—canals, glands, and nerves with otherworldly names like the glands of Zeis, or the crypts of Henle. The lachrymal gland itself produces most of the tears we shed when our eyes are irritated or we feel our deepest emotions. Basal tears drip continuously from a system of tiny glands at the top of the eye and combine with fluid from goblet cells, the glands of Manz, and forty-six additional glands to create a complex plumbing system that keeps our vision clear and our eyes free of disease.

Most tears eventually find their way to the canals at the bottom of our eyes near the bridge of our nose, where they then drain through the puncta, the lip of tissue at the edge of our eyes. From there they journey through the lacrimal sac, past the valve of Hasner, and into our nose, which explains why crying hard leads to runny noses.

The whole system, however, can handle only so much water. It empties tears at the paltry rate of a microliter and a half per minute, a droplet only slightly larger than the tip of a ballpoint pen. If we cry heavily, the system overflows and our tears well up and spill down our cheeks. As it turns out, this is extremely important because visible tears are crucial to human communication.

Scientists categorize the kinds of tears we cry not only according to their purpose or the events that cause them, but also by their chemical makeup. Reflex tears, like basal tears, come loaded with globins and glucose, antibacterial and immunological proteins, urea, and lots of salt. But emotional tears have a different chemical makeup. In fact, the tears we cry when we are upset, as opposed to when we get a sharp stick in the eye, have 20 to 25 percent more proteins. They also have four times the amount of potassium normally found in blood plasma, and thirty times the concentration of manganese. Psychic tears brim with hormones—adrenocorticotropic hormone

(ACTH), for example, an extremely accurate indicator of stress, and pro-lactin, which controls the neurotransmitter receptors in the lacrimal glands that release tears in the first place. The strange thing is that prolactin is also the hormone that makes it possible for women to produce breast milk.

Scientists believe that these cocktails of hormones and proteins are linked to the moods, stresses, and emotions we often associate with crying. High concentrations of manganese, for example, show up in the brains of people suffering from chronic depression. Too much ACTH is an excellent indica-tor of increased anxiety and stress. And studies show that women—all of whom have higher levels of prolactin than men—cry about five times as of-ten. In fact, women unfortunate enough to have inordinately high levels of prolactin also experience more intense hostility, anxiety, and depression, which can lead to still more crying.

There's another perplexing connection between prolactin and tears: when the child of a breast-feeding mother cries, the mother's milk reflexively "lets down," so that it becomes immediately available to the baby. In other words, the mother's body instantly and reflexively prepares itself to relieve the in-fant's reason for crying, or at least the most likely one. Some mothers even claim to experience a kind of telepathy with their infants, times when their milk lets down even when they are traveling or at the office running a meet-ing miles away from their infants. They report that on checking in later, their milk let down at precisely the time their baby started to cry.[4]

Just as different reasons for crying result in chemically different tears, different parts of the brain—the seat of our experience—are physically connected to the feelings that make us cry. Nerves linked to our lacrimal glands rope their way by long, roundabout routes into both ancient and newly evolved areas of the brain—the pons, basal ganglia, thalamus, hypo-thalamus, and prefrontal cortex. Each of these areas are themselves im-portant cerebral routing stations that handle functions and experiences as diverse as facial expression, breathing, body temperature, sight, swallowing, and reflection, memory, planning, and worrying. So it is no accident that so many kinds of feelings can cause us to cry. Nor is it an accident that those feelings affect our body temperature and blood pressure, heart rate and fa-cial expressions, at the very same time as they evoke more memories and emotions—many of them confusing and contradictory.

When we cry, some of the hormonal cocktails that are driving the feelings we experience actually find their way into our tears. Biochemist William

Frey, who directs the Dry Eye and Tear Research Center in Minneapolis, believes that one of the reasons why we feel better after we weep is because we are literally crying out the extra hormones and proteins in our brains that generate the feelings that saddened us in the first place. This explains why, he says, we sometimes advise one another to "Go ahead. Have a good cry." Emotional tears are the body's way of flushing out the chemicals that make us sad—the excess prolactin, manganese, and ACTH.

Crying Factoids

Though crying is poorly researched, a few studies have yielded some interesting statistics. For example, do women cry more often than men? Dr. Frey and his colleagues worked with 331 volunteers, aged eighteen to seventy-five, and asked them to keep a "tear diary" for thirty days. Women reported crying four or five times more than men that month. Frey theorizes the reasons are more chemical than cultural. Women have much higher levels of serum prolactin than men, and prolactin is a hormone connected with the production of tears as well as breast milk. "Hormones may help regulate tear production and have something to do with crying frequency." The fact that there is very little difference in the rates of crying among boys and girls bolsters Frey's argument. Until about age twelve prolactin levels are about the same no matter the sex, but between the ages twelve and eighteen women develop levels 60 percent higher than men and cry more often.

Based on Dr. Frey's "tear diaries," participants reported that they cried because they were sad 49 percent of the time; out of happiness 21 percent of the time; in anger or sympathy, 10 and 7 percent, respectively; and because of anxiety or fear, 5 and 4 percent, respectively. The remaining reasons for crying were unaccounted for.

According to another study, a happy cry averages two minutes; a sad cry, seven.

Not everyone agrees with the theory that emotional tears drain out the hormones and proteins that made us sad. If, for example, you find yourself crying over the death of a close friend as you recall the good times you shared, scientists wonder if it is your memories generating the hormones that make you sad, or the other way around? There is no absolute way to know. It may be both since, in effect, the brain is a tremendously complex feedback loop (hundreds of millions of loops within loops, really) that continually interact with the world outside and its own kaleidoscopic experiences inside. Perhaps feelings generate hormones, and hormones generate more intense feelings until we finally burst into tears.

Randolph Cornelius, a psychologist at Vassar College, has been investigating the profound emotions that emerge out of crying's intricate neuronal alchemy since he began researching his doctoral dissertation some twenty-five years ago. During the earliest days of his work he made one simple request of the subjects he was studying: Tell me about the last time you cried in front of another person. The heartbreaking and harrowing accounts of life at its extremes often brought him to tears himself at the end of the day.

There was a young woman, for example, who told him of the day when she was barely nineteen, standing in a hospital with her six-month-old infant in her arms, and gave doctors permission to remove life support for her husband, who was dying of cancer and had suffered a heart attack. She held her tears, she told Cornelius, until she had signed the papers, and then fell apart in the arms of a nurse whose name she never knew.

During another interview a Vietnam veteran told how he had had half of his face shot away during a firefight in Vietnam. He lost an eye and had a metal plate implanted to replace the portions of his skull that had been shattered. One day, he told Cornelius, he made a call to his therapist. He planned to leave a message telling her that he was committing suicide. She unexpectedly answered the phone, and in the conversation that followed, he had a major breakthrough. When he did, he told Cornelius, he could feel "hot tears rolling down my cheeks." But the thing was, he said, he knew he was missing the eye from which he felt tears were falling.

A powerful sense of empathy and sympathy is a uniquely human reaction to the tears of others. Crying often begets more crying, possibly because the mirror neurons that long ago made it possible for our ancestors to learn how to make a tool also enable us today to put ourselves into one another's emotional shoes.

This makes human crying an unusually potent form of communication. More than anything, tears reveal us at our most vulnerable. Laughter bonds us, and it can bring us progressively closer, but crying binds us in another, deeper way: It is an unmistakable plea for help, an expression of utter vulnerability. Tears in some intense and commanding way communicate an opportunity for intimacy and authenticity that is more powerful than any words could be. When we cry, the walls are down, and the defenses have been breached.

. . .

Whatever the reasons why we cry, it is difficult to argue that tears alone can sufficiently flush the hormones from our bodies to provide the sense of relief we feel after crying. Our tear ducts simply aren't that big or that efficient. Even a good, long, heaving bout of crying doesn't add up to more than a thimbleful of hormone-laden fluid.

Yet it is almost always true that we *do* feel better after we cry, even if only temporarily. When dealing with terribly sad situations, a death or the loss of a relationship for example, crying at least seems to bring us a breather, a chance to regroup emotionally. But if we don't feel better because we have flushed our systems of the chemistry that causes our sadness, what's the source of the relief?

One explanation is that we may not only cry hormones out, we also may cry them in. Frey also has discovered that the neurotransmitters leucine-enkephalin (a natural opiatelike substance that relieves pain) is released in the brain when we weep. These neurotransmitters are like the ones that laughing generates, though obviously for different reasons. But the effect is similar: They improve our mood. At first blush it is a mystery why this should make good evolutionary policy. After all, what purpose would mood swings serve for an animal when it comes to surviving predators and disease? Not much, unless that animal happens to be as intensely social and intelligent as we are.

In nature, it is generally good to maintain a state of homeostasis—to remain neither too hot nor too cold, too active nor too lethargic. And if there have to be swings in one direction or another, it is best to at least keep them under control and get back to normal as quickly as possible. If open living systems including bacteria, trees, ocean reefs, and humans fail to rectify extremes that pull them too far from their normal style of living, they will, and do, unravel and die. A plant may freeze, a lizard may overheat, a forest may be denuded, and a human might become so emotionally undone he can't function effectively.

In a species whose survival depends on maintaining stable relationships, equilibrium embraces more than just physical comfort, it also embraces psychic comfort. Our need to maintain homeostasis explains why we eat and sleep, why we invented shelter and clothing and air conditioning, and it may help explain why we cry.

We all learned in our grade school science classes that the autonomic nervous system controls "mindless" operations such as breathing and, heartbeat as well as the basic functioning of our kidneys and brains. But the autonomic nervous system is itself divided into two other systems: the sympathetic and the parasympathetic. The sympathetic nervous system evolved to prepare us for action—physically, mentally, and emotionally. When we are scared, for example, the sympathetic nervous system fires off the messages that tell our bodies to skedaddle or stand our ground and prepare to fight.

The traditional thinking for years was that because the sympathetic system gets us emotionally excited, it also must be the system that causes us to cry. But now many scientists think it may be the other way around. After all, after every fight or flight, we have to settle down. If we continued to run in overdrive, we would blow an aorta, or have a stroke, and that would be the end of that. Given the dangerous nature of the lives our ancestors lived, it wouldn't have been long before the species would have been done in by cerebrovascular accidents or coronary thrombosis. So the parasympathetic nervous system returns our neurotransmitters, heart rate, and hormones to normal. Quite possibly we cry not because we are getting agitated and upset, but because it is a way for our nervous system to bring us back into equilibrium.

One study reveals, for example, that if the nerves central to the sympathetic system are paralyzed, patients cry more. But when important parasympathetic nerves are damaged, they cry less. If crying was driven by the sympathetic nervous system, it would be the other way around. In other words, we don't cry because we are upset, which is the way it feels, but because we are trying to get *over* being upset. That may be the real reason why we feel better after we have a good cry.

Looked at this way, it is easier to see why crying might have evolved. Like so much that evolution favors, crying is a survival strategy, like eating food or sleeping or breathing air, all things we do to stabilize, get into our comfort zones, and stay alive.

Still, none of this explains why we cry tears. We could just as easily, and tearlessly, howl like a coyote or scream at the top of our lungs like our chimp cousins, and still feel some relief. But where are the evolutionary advantages in tears? After all, they blur our vision and add to the vulnerability our scrambled emotions have already created. Who among us would feel comfortable knowing that the pilot flying our coast-to-coast jet, or the doctor performing brain surgery, was in tears? Yet somehow they must help; otherwise the laws of evolution, would have booted them from the gene pool long ago.

In 1975 an Israeli biologist, Amotz Zahavi, conceived an interesting theory about why animals behave—on the surface, at least—in ways that don't seem to have much evolutionary purpose, but on closer observation turn out to be perfectly sensible. Many of these behaviors, he pointed out, seem not only baffling, but often downright counterproductive. Why does a peacock have enormous, colorful tail feathers when the natural result of having them must be that it slows the bird down, draws the attention of predators, and makes it difficult to fly? Or why does a gazelle, when it senses a lion is about to attack, bound straight up into the air like a pogo stick before making its exit?[5]

Zahavi calls these traits and behaviors examples of the "handicap principle." We see them everywhere in the natural world, from the huge antlers of bull elks to the loud squawking of hungry baby birds. On the surface the traits make no sense for the simple reason that they come at a high price— they require energy and resources, and they draw dangerous amounts of attention.

But according to Zahavi they also serve as powerful forms of communication. In fact, the bigger the handicap, the more powerful the communication. An antelope's first vertical bound, for example, immediately puts it at a disadvantage. It has lost precious seconds it could have used to put distance between it and the predator that intends to make a meal of it. But a leap like that also sends a message that says, "I am so fast and can leap so high, there is no chance you are going to catch me. So don't waste your energy." Often the lion or cheetah poised for pursuit absorbs the message, performs a quick cost-benefit analysis, and walks away in search of more sickly prey incapable of a five-foot vertical leap.

In a strange way, messages like this have introduced a kind of primal form of truth and honesty into the natural world. They can't be faked because they are too costly. If a peacock could fake big, heavy tail feathers and wasn't truly healthy enough to support them, it would quickly find itself a meal for any fox or wildcat that called its bluff. Those genes would not be passed along. The same would be the case if an antelope were capable of faking one bound, but was then unable to head off in the opposite direction at lightning speed. All of this is nature's way of saying "If you can't walk the walk, don't talk the talk."[6]

Being "truthful" may explain the origins of our tears. Crying, like laughing, is a unique form of human communication, and its roots are just as primal. But unlike laughing, which we all do all the time, crying is reserved for special occasions. That we don't often cry tears of emotion indicates that they come at a cost, a Zahavian handicap that requires extra energy or invites undue attention. Because tears are costly and rare, and because they are cried only when very deep emotions are being felt, they are not easily faked and send an unmistakable signal that the feelings behind them are absolutely real.

More often than not, tears signal that we want help and consolation because we are in pain—physical or emotional. But it also applies to feelings of pride or joy. When a father cries at the sight of his newborn baby and his wife sees this, it bonds them and says, in effect, "We are in this together." Cornelius has observed that our response to a person weeping tends to be profound, and reinforces the connections we feel for one another. Our hearts go out to people, even total strangers with whom we have nothing in common.

This cuts both ways, however. When we see someone crying, but we don't see any tears, we are immediately suspicious. Tearless crying simply isn't as authentic. Cornelius tested the dynamics of this, too. Over the past six years he and his students have been gathering still photographs and video images from newsmagazines and television programs, all of them of people crying real and visible tears. When they found a particularly appropriate image, they prepared two versions: one, the original, with tears; and another, with the tears digitally erased.

For this experiment they gathered a group of people and sat them down one at a time in front of a computer monitor to watch a slide show. Each slide presented two pictures: one tearful, the other a different picture with the tears secretly erased. No participants were allowed to see the same picture with *and* without tears. Cornelius's team then asked each participant to explain what emotion they thought the person in each photograph was experiencing, and how they would respond to a person with that particular look on their face.

The test's observers universally registered that pictures of people with tears in their eyes or on their cheeks were feeling and expressing deeper emotions—mostly sadness, grief, and mourning—than those who were tearless. But when participants looked at pictures where the tears had been digitally removed, they often concluded that the people in them were feeling any number of emotions; from sadness to awe to boredom. Cornelius concluded that tears alone sent a specific and powerful emotional message.

But there was another wrinkle. The tearful pictures often elicited two very different responses from those who looked at them. About half of the observers felt people crying tears were saying "Help me" or "Comfort me." The other half felt that the people in the images wanted to be left alone. This seems, says Cornelius, to have less to do with the expressions on the faces in the pictures and more to do with the attitude of the people looking at them. Sometimes crying is a call for help and attention, but sometimes it is a way of saying we are vulnerable and want room and space to deal with what is upsetting us until we can get back to normal. A contradiction? Not necessarily, says Cornelius. Trying to hide our tears still sends an honest signal about our state of mind. It still says we are in trouble, even want comfort, just not necessarily from someone else right at that time. Paradoxically, if someone refuses comfort or tries to hide their crying, it makes them seem even more vulnerable.

Tears carry a lot of expressive weight. If we evolved our outsized brains mainly to handle the complex social and personal interactions to which we pay tireless attention, then tears add one more true and powerful arrow to the quiver from which we draw our many forms of human communication.

Dr. Paul MacLean, chief of the National Institute of Mental Health's Laboratory of Brain Evolution and Behavior, has speculated that human crying has its origins in distress and separation calls similar to those of young primate monkeys and chimps. This means that like laughter, the *sounds* of crying have their roots in the hoots and howls of the jungle. He has speculated that in the course of evolution, the early use of fire and the smoke that rose from funeral pyres irritated our ancestors' eyes, and in time became associated with sadness. It is a theory, and obviously our lacrimal glands somehow wired themselves into the emotional centers of our brain. But the truth is there is no way of knowing how precisely.

Evolution operates by random chance, after all. A genetic mutation enters the world packaged in the creature that exhibits it, whether it is colorful feathers or tears. If that creature survives, the adaptation does, too, and is passed along. And if it works, it spreads out into the next generation and so on. Crying, as MacLean has theorized, may well trace its origins to the hoots and calls of our precursors. Countless species come into the world squawking or screaming for attention. Perhaps there is some clue here into how primal calls somehow morphed into the tears we cry.

. . .

When we enter the world, our loud bawl of a birth cry sends the indisputable message that we have arrived. Later cries become more sophisticated: In babies, they communicate hunger, pain, loneliness, and discomfort. During the first eight months of life, human infants don't actually cry tears. They don't have the plumbing for it. Nor is it necessary in those early months, because they are so clearly helpless that every cry is believed.

But in toddlerhood the situation changes. Crying becomes subtler, a kind of simple language, and it can be used to manipulate. After all, children, even as they grow older, want the attention of their parents, and since crying has been their most effective way of getting it, they continue to use it, even when they don't absolutely need help. The babies of rhesus macaque monkeys even behave this way. They cry out to their mothers in infancy, and tend to cry even more at about the time their mothers wean them. (Macaque monkeys can only become pregnant again when they stop breastfeeding. Contrary to popular belief, this is *not* true of human mothers.) At first macaque mothers come running, but as the cries increase, they begin to respond less because so many turn out to be false alarms. The macaque moms become more skeptical, and eventually the infants cry less because it isn't working.[7]

Crying wolf like this, however, also may have made tears even more effective. Every parent has experienced the tearless crying of a child who is unhappy and wants attention but isn't really in deep trouble (the scientific term for this is whining).[8] So among the first signals parents look for when a child cries are real tears, a sure sign that their toddler truly needs help (as opposed to just wanting a bag of Snickers at the grocery store). Presumably the same would have been true for our ancestors. Tears would have put the same visual exclamation point on any howl, moan, or grimace for them as much then as they do today for us.

At some unknown and unknowable time the gap between the reflexive tears that a poke in the eye creates and the emotional tears a broken heart brings on closed. During the past six million years, enormous changes took place in our ancestors, much of it from the neck up. Not only have our brains doubled in size and then doubled again, but also our faces have changed and so have our ways of conveying emotion with them. Their rich, expressive musculature evolved by chance, but remained with us because it helped us more precisely communicate, and sometimes manipulate, one another.

This may explain how tearful, human crying evolved. Somewhere at some time the genes that control the parts of the brain associated with the experience and expression of emotion became connected, literally, to the lacrimal gland that sits above each of our eyes. This would have been a first in the primate world. Chimps and gorillas will grimace and howl, snarl and moan. They feel sad or hurt or scared, but they cannot cry tears of pain or frustration or joy any more than they can stand up and walk around erect as an English butler or begin talking in full sentences.

Perhaps sometime long ago, when our ancestors were still living chimp-like in Africa's rain forests, one was born with the odd ability to cry tears when she was upset, but because it didn't serve any useful purpose in her social world, the gene eventually disappeared. The personal interactions of jungle-living chimps are highly complex, but not nearly as complex as ours. Tears might have amounted to communication overkill, just as evolving the many facial muscles we have today would have been.

But for our kind the situation was different. Our ancestors lived outside the relative safety of the jungle. And because they did, they needed one another more than their forestbound cousins. And if they needed one another more, then they also needed to communicate more effectively. So as their brains and social interactions grew progressively more multifaceted, one feeding on the other, their need to bond and communicate, manipulate and read one another's minds also accelerated. Complex relationships beg for increasingly complex minds and increasingly complex forms of communication. Language was one mighty adaptation that emerged for this very reason.

Tears, given the potent, highly visible messages they send, were another. They became a kind of supreme form of body language.

But wouldn't speech have sufficed to express our strongest emotions? Wouldn't the precision of words trumped crying, even tearful crying? Maybe, except that truly strong emotions often linger beyond the reach of words, and tears do what syntax and syllables can't. We all know the feeling, whether it is profound sadness, frustration, anger, pride, or pain. Crying expresses emotions that words simply can't.

Maybe it is not so much that the lacrimal glands of chimps are not wired into their brains (because they are), but that there is no oversized neocortex in there to which it can make a connection. Plato wrote twenty-five hundred years ago that our natures are driven, like a chariot, by two horses, one dark and wild (our emotions), and the other controlled and logical (our intellect). Intellect, he said, must control our dark side. And in some ways, the evolution of the prefrontal cortex supports that simile. But it is more complicated than that. Our intellect doesn't simply suppress our visceral, animal feelings. It also supercharges them. And maybe this is the central reason why crying is a uniquely human trait. It marries raw emotion with a brain capable of reflecting on those howling, primal feelings. That is why we cry. Our simian cousins, gifted and intelligent as they are, don't have the capacity for the powerful marriage of thoughts and emotions. They can feel rage, frustration, or loss, but they do not reflect on them. The random emergence of genes that connected the emotional *and* intellectual parts of our brains to lacrimal glands that sit above our eyes gave us a new way to express those elusive feelings. And in the bargain we gained an emotional stamp we can put on our cries for help that no other creatures possess.

VI
Kissing

Chapter 10

The Language of Lips

kisses are a better fate than wisdom.
—e. e. cummings

If you are ever in doubt as to whether to kiss a pretty girl, always give
her the benefit of the doubt.
—Thomas Carlyle

Kisses are like tears; the only real ones are the ones you can't hold back.
—Anonymous

WHO DOESN'T CARE FOR A FETCHING SMOOCH, a long-lasting liplock, a lingering kiss? We have an affection for kissing because our lips enjoy the thinnest layer of skin on the human body, and the nerve endings there, and in our tongues and mouths, send messages to the brain that define the word "pleasure." This explains why the brain devotes great swathes of real estate to the nerves needed to operate these parts of our bodies—more, even, than it devotes to moving our entire torso. No part of our anatomy is better tuned to the things that touch them. Lips, it seems, are all about sensation. And so we kiss—furtively, lasciviously, shyly, hungrily, and exuberantly. We give ceremonial kisses, affectionate kisses, social kisses, kisses of death, and kisses that give life. When passion takes a grip, we lock together exchanging not just fluids and breath, scents, and tastes but souls and minds, feelings, secrets, and emotions that defy words and run circles around syntax. It is as though a circuit has been completed and the currents of two hearts have directly fused into something entirely new, which, in a sense, they have.

You are unlikely to notice when you are locked in a passionate, soul-searching kiss, but your pulse rate and blood pressure are rising, your pupils are dilating, and you are breathing (when you can get a breath) more deeply.[1] Your kissing is also reducing your chances of tooth decay, relieving stress, burning calories, and increasing your self-esteem.[2] Part of the reason your self-esteem is rising is because lips meeting lips releases waves of the neurotransmitters norepinephrine, dopamine, and phenylethylamine (PEA), which in turn attach to the pleasure receptors in your brain that generate the same feelings of euphoria we feel when we laugh or exercise hard or take certain mood-enhancing drugs such as cocaine or heroin. This is why you will almost never be depressed when kissing.

Kissing also requires your face to work hard. According to Margaret H. Harter, at the Kinsey Institute for Research in Sex, Gender, and Reproduction, even when we pucker up for nothing more than a standard hello or good-bye, thirty lip muscles go to work.[3] While they do, neural connections that run from the lips, tongue, cheek, and nose up to the brain enable kissers to sense temperature, taste, smell, and movement, which in turn drive the production of those pleasure-making neurotransmitters. Of the twelve cranial nerves that affect brain function, five are in play when we kiss. If we each generate our own personal weather, you might say kissing makes an excellent barometer for detecting it.

Osculation (the scientific term for kissing) is not the reason why we developed all of these nerves and muscles. They originally evolved for eating, and the forces of natural selection have refined them to sense flavors and textures and tastes that can make the difference between a nice snack or an ugly, poisonous death. Some scientists wonder if this might explain why we are the only mammals with red, outwardly turned lips. But others, like Desmond Morris, theorize that just as the red and blue faces of male mandrills are facial imitations of the sexual messages their colorful rumps send, our red, pouting lips recapitulate a woman's genital labia. Like sexual labia, after all, our lips (women's more than men's because women's are fleshier) become redder and more swollen when we are sexually aroused and more blood flows their way.[4]

This swelling is itself arousing, which explains the wealth of long-standing cultural evidence that men find it attractive, and women often do what they can to enhance it. Egyptian ladies painted their lips a reddish purple with a plant dye called fucus-algin.[5] And seventeenth-century European women rouged their lips enough that Thomas Hall, an English pastor, felt compelled to write that those who did were brazenly trying to "kindle a fire and flame of lust in the hearts of those who cast their eyes upon them."[6] Today lips continue to be enhanced, and at a record pace. Lipstick is a $1.5 billion business, and among the more popular current cosmetic procedures is a silicone injection designed to make every pair of lips appear as though they belong to Angelina Jolie. Some psychologists even believe that pouting and lip biting are forms of body language that developed to puff lips up, all in the primal interest of getting the two sexes together to make more versions of ourselves.

However our lips came to have their unique shape and special sensory capabilities, in kissing they found new purpose. We use them to send messages far more dramatic than any words they can form, and nourish hungers deeper than any food mouths consume. As actress Ingrid Bergman once observed, "A kiss is a lovely trick designed by nature to stop speech when words become superfluous." A "secret told to the mouth instead of to the ear" is the way Edmond Rostand put it.

If we look at it this way, kissing is another human variation on the thing we do better than any other species: communicate. Like laughter and crying, it reaches deep into our past to tie old parts of our nature up with new ones and create behavior that only we are capable of. When we kiss, our history and evolution—all of the wheels and gears and chemistry that make us go—are plastered all over the whole tender, tempestuous, and splendidly human act.

. . .

Kissing is not a universal human behavior. About 90 percent of us do it. But that means some 650 million of us don't, more than the populations of every country on the planet except for China and India.

Why this is the case is hard to fathom. It is such a sumptuous, luscious invention, you would think it is as deeply spun into our DNA as breathing and walking. But it is a cultural, not a genetic innovation, which is to say we are not born kissers. We have had to learn it.

Back at the turn of the twentieth century, for example, Danish philologist Kristoffer Nyrop found that members of certain Finnish tribes bathed together completely nude, but considered kissing indecent. In Mongolia, some fathers still do not kiss their sons. Instead they smell their heads. An "Eskimo kiss" involves rubbing noses, not touching lips. Neither Polynesians nor Maoris prefer to show affection by kissing. In 1897 the French anthropologist Paul d'Enjoy (an apt name for a budding philematologist—a scientist who studies kissing) reported that the Chinese found mouth-to-mouth kissing nearly as horrifying as we find cannibalism. When Charles Darwin first visited the inhabitants of Malaya, he reported that there was absolutely no osculation going on, but plenty of nose rubbing. And the explorer Captain James Cook discovered similar behavior when he first visited Tahiti, Samoa, and Hawaii.[7]

Kissing may not be for everybody, but apparently it never fails to become popular wherever it is introduced. By all accounts, once Cook's crews made landfall, everyone quickly became amateur philematologists. And today kissing is common throughout China. It is strange to say in an age of cell phones, computers, and satellites that we may actually be witnessing the last evolutionary stages of a cultural behavior that has been sweeping through the race for tens of thousands of years but hasn't quite reached the finish line.

But even if all of us haven't yet quite learned to kiss, how did any of us get started in the first place? Pheromones may provide a clue. In 1995 an enterprising group of Swiss researchers headed by zoologist Claus Wedekind decided to test the effects of scents on human behavior. The idea was that certain mysterious clusters of molecules called pheromones might shape some very important, even life-changing decisions that we all make, without us even consciously realizing it—whom we choose to marry, for instance.

Wedekind and his team first gathered up forty-four men and forty-nine women. They next tested and profiled all of their immune systems to map which diseases they would tend to resist well and which they wouldn't. Wedekind then instructed the men to wear the same T-shirt to bed for two consecutive nights. Each was given unscented soap and told in no uncertain terms not to tamper in any way with how he smelled. Off went the men.

At the end of the obligatory two days, the same forty-four fellows took off their well-slept-in shirts and deposited them into several boxes that also contained other, completely new T-shirts. Now the forty-nine women were assembled, asked to sniff the shirts, and questioned about which ones they found most "sexy." You would think the universal answer would be a resounding "None of them!" But as it turned out, the women did have a preference—and strangely enough, the shirts they preferred belonged to men whose immune systems were dramatically different from theirs. The women, in other words, tended to be attracted to the men who, if they were to have children with them, would father offspring capable of fighting off more diseases than either of their parents. That, of course, would make them more likely to survive (and pass along *their* genes).[8] A very primal chemistry seemed to be at work.

The standard definition of pheromones is that they are naturally occurring compounds that instigate some remarkable behavior in members of the opposite sex. Scientists have known for some time that they exist in the insect and animal worlds, where their effect is indisputably powerful. For example, the cecropia moth, the largest in North America (and one of the most beautiful), has been known to catch a whiff of a female's pheromones and gamely push itself upwind a full seven miles to reach and mate with her. Social insects such as bees, wasps, and ants can't live without pheromones. They use them to maintain the complicated societies in which they live and work, which explains why their antennae insistently search out and wave at their surroundings wherever they go, hunting invisible molecular messages that signal them what their next move should be.

Mammals also rely on these chemical communicators. Male pigs exude a pheromone that goes by the nondescript name 5-androst-16-en-3-one, and it makes sows as amorous as Mae West in *Night After Night*. It is so reliable, in fact, that it is sold as BOARMATE to pig farmers who need to keep their supply of new piglets uninterrupted.[9]

Though the list of pheromone-driven behavior in nature is long and fascinating, scientists continue to wrangle over the precise role it plays when it comes to humans. For years they were thought to have no place in the human world. But the evidence is mounting, as Wedekind's experiment indicates, that pheromones have a good deal to say about how we behave, especially when we are in the presence of the opposite sex.

In the Swiss test, it was as though certain chemical messengers evolved to ensure that opposites attract, at least opposite immune systems, the better to maximize the continuation of the species. You can see how this might be useful out on the savanna, where the survival of the group would take precedence over personal likes and dislikes. Without this kind of chemistry our ancestors may have gone extinct before they became our ancestors.

If pheromones do play a role in our personal lives, it might mean that today the reasons why we find some people appealing and others not has nothing to do with why we *think* we find them that way. Love, or at least attraction, might be blinder than we ever suspected. Whatever the case, our pheromonal predilections apparently go way back. We may not only share the trait with pigs, bees, and moths, but also with rats, who "read" the pheromones in the urine of the opposite sex and choose their mates accordingly, a kind of chemical date-matching system.

The difference between us and other mammals, however, is that they search for pheromones actively, sniffing out the mates most likely to create strong offspring and avoiding those who won't. We, on the other hand, are completely clueless that any of these chemical conversations are taking place. The communication is entirely unconscious, which may help explain why some people fall in love "at first sight." Love at first smell may be more accurate.

Until relatively recently it would have been unthinkable in scientific circles to theorize that silent chemical envoys secretly affect intimate decisions such as choosing a suitor or a spouse. Insofar as scientists thought about these questions at all, they had concluded that we simply didn't have the organs necessary for sensing such things. Pound for pound, after all, rodents have much larger olfactory bulbs than we do. The chemistry of smelling is a much larger part of their world, and the worlds of most other mammals, including dogs and cats and cows. Part of the fragrance-sensing apparatus in rats, for example, is something called a vomeronasal organ, or VNO, which has evolved just for the recognition of pheromones. The VNO senses molecules in the environment that accelerate puberty, reveal pregnancy (and failed pregnancies), and even generate testosterone surges. It tells male rodents when it is best to mate and signals them when it is a bad idea.

For years it was assumed that if we humans ever had organs like these, we would have forsaken them long ago for lack of need. After all, why send messages chemically when we have these big brains and lavish vocabularies? But then in the 1970s, in a still-famous experiment, Martha McClintock, a psychology student at Wellesley College in Massachusetts, studied the menstrual cycles of 135 women who lived together in her college dormitory. She found that after spending several weeks in one another's company, the menstrual cycles of her female dormitory mates drew together in nearly perfect synchrony. In other words, it appeared that bodies that had come together from all around the country to live in the same place had begun talking with one another, coming to agreement, silently and chemically, for no other reason than that they were in one another's company. How?

Obviously this synchronization wasn't the result of any conscious effort. Women cannot purposefully control the timing of their hormonal cycles. And though McClintock suspected that pheromones were behind the synchronization when she published her now famous findings in *Nature* in 1971, she couldn't say so for certain.[10] The technology didn't exist for such minute molecular measurements. Nevertheless, her findings indicated that to the long list of complex ways we communicate, we could apparently add that we connected with one another by a kind of chemical telepathy that had nothing to do with conscious thoughts or feelings.

Following McClintock's study, more evidence for human pheromonal communication emerged. In 1985 researchers at the University of Colorado discovered that we did indeed have working vomeronasal organs after all, in the form of a very small pair of ducts just inside our noses, one tiny pit in each nostril. The research even turned up evidence that the human VNO had its own proprietary nerve connections, wired directly to the brain, that operated altogether independently of the olfactory system that does most of our nose's work. That meant that strictly speaking, pheromones didn't really have anything to do with odors because we don't actually smell them (although we may smell other odors that accompany them). Apparently it is all about the molecules reaching the VNO and triggering parts of the brain that in turn trigger very specific behavior. This makes pheromones the "words" in a direct, molecular conversation between the brain of one person and the body of another.

Women, for example, seem to be particularly tuned into a male pheromone called androstenol, an odorless molecule produced by male sweat glands. When women see pictures of men that have been sprayed with androstenol, they actually find the picture more appealing than one that hasn't been sprayed, and it doesn't matter whether it is an image of Mel Gibson or Joe the plumber. Another study, at University College in London, revealed that when women were exposed to androstenol they were more likely to have greater social contact with men in the hours that followed. It may not be BOARMATE, but it apparently made them more outgoing than usual, but only when they encountered men.

Still another study revealed that when theater seats were randomly inoculated with androstenol, women were more likely to sit in them than seats that were pheromone-free, unaware that they had been led to the seat literally by their noses. An experiment at the Monell Chemical Senses Center in Philadelphia even found that when women are consistently in the company of men, the length and timing of their menstrual cycles changes. They tended to ovulate more regularly, basically making them more predictably fertile. For mating couples who want to bring more babies into the world, that would be a good thing. But how odd that such personal behaviors are affected by chemicals about which we are completely oblivious.

Pheromones can have strange effects on men, too. One recent experiment showed that the female hormone copulin seemed to tune men into the women around them who were "wearing" it. To test a similar idea, the ABC News show *20/20* conducted an informal test. The network recruited two sets of twins, one male and one female. A member of each set was then sprayed with an unscented pheromone that was supposed to attract the attention of the opposite sex. The other members got nothing more than plain old witch hazel. They then stationed themselves in separate sections of a popular New York City bar, where they were told to just sit and do nothing. No charming smiles or evocative body language were allowed.

Neither of the male twins saw much action. They both received about the same amount of attention from women in the bar as they always did. But for the pheromone-scented sister, it was a different story. She attracted thirty different men, nearly three times more than the eleven whom her equally attractive twin received. The only difference, apparently: the silently communicating pheromone she had been sprayed with.[11]

How this happens precisely is still a mystery, but it provides some clues to pheromones' connection to kissing. They seem to have a magnetic effect, encouraging members of the opposite sex (and gay members of the same sex[12]) to want to cuddle and embrace. We have all seen animals nuzzling, perhaps because their pheromones are at work. Some scientists have wondered if the nose rubbing of Inuits and Malayans is a nuzzling precursor of kissing, an effort to get closer to one another so we can more deeply inhale the intoxicating messages are bodies are tossing off.

So far the best that scientists can say is that the human VNO provides a hotline to our hypothalamus, the part of the brain that affects everything from body temperature and heart rate to emotions and sexual appetite. It is the hypothalamus that then goes on to revamp our behavior.[13]

Imagine the effect this might have had among our precursors—early humans throwing off chemical signals like party confetti that drew or repelled one another, totally unaware of how it was all happening. Picture *Homo erectus*, Neanderthal, or Cro-Magnon couples, sniffing, rubbing faces, and cuddling, mingling their scents. Unlike in the love life of rodents, these pheromones would not have been the sole arbiters of affection and attraction, just silent partners that led to the first kiss. As cheeks and breath and roaming hands came together; as personal attractions converged with primal chemistry and sensations grew stronger and richer, kissing—literally tasting and sharing one another with the most sensitive and sensual parts of our bodies—would have been a perfectly natural next step.

In fact, new research has revealed that while we sense pheromones (and all of the feelings they can induce) by hijacking them out of the air, we also can "taste" pheromones with our tongues and lips, which enables us to gather even more chemical information from the person we are kissing. In other words, osculation would have been a natural, if unconscious, way to scan the gene pools of potential mates even more completely than simply inhaling. Maybe that is another reason why nature ensures that kisses feel so good. It is a way to encourage us to find the best possible mate so we will create more human beings, the ultimate goal of every self-respecting strand of DNA.

. . .

The driving search for pheromones is not the only current theory about why we began to kiss. It is also possible that it began as a prehistoric feeding practice between parent and child. It could be that the two apparently separate theories are actually connected. The idea behind the feeding hypothesis is that hominid mothers prechewed food for their young, not unlike birds and other animals do, and then passed it along their children in a kind of kiss. Chimpanzees do this. Pressing outturned lips against lips may then have later developed to comfort a hungry child when food was scarce, and eventually come to express love and affection. In time this parental kissing might then have found its way into use among lovers who brought new variations to the practice.

There is a certain amount of sense in connecting nurturing and appetites with kissing. It introduces emotion into the equation, and it makes kissing more than a soulless fishing expedition for finding fertile, healthy mates.[14] Food, hunger, passion—they are all connected on some level. An ancient poem from the Bible's Song of Songs (4:11) says it all: "Thy lips drip as the honeycomb, my spouse:/Honey and milk are under thy tongue." We are all hungry for affection, after all, and for someone out there.

Sustenance and kissing have even been tied up in the linguistics of ancient Egypt. For a while Egyptologists mistakenly translated the hieroglyph for "eat" as the word for "kiss," but kiss fit so well in sentences that involved eating that it took years for them to realize they had it wrong. Looking at it this way, it is not so difficult to see how our kissing may have migrated from satisfying one form of hunger to satisfying others.

. . .

Despite its pleasures, kissing does entail some risks. By one count a single meeting of the lips can exchange 278 species of bacteria and viruses, some good, some not so good.[15] Colds, flu, bubonic plague—all can lurk behind a loving buss. But on balance, kissing seems mostly to have helped our kind survive because it leads so pleasurably and naturally to sex, and sex, rather inevitably, leads to babies.

This path has been, of course, problematic for the millions of teenagers over the years who have been shocked to suddenly find themselves parents. The onset of puberty, and the oceans of hormones that go with it, evolved long before modern culture did, and they drive behaviors today that once made better sense in a world where fourteen-year-old mothers and fifteen-year-old fathers were more the norm than the exception.

In those days the emphasis was less on developing deep and emotionally abiding relationships, and far more focused on the survival of the species. Though there were surely prehistoric examples of warm primate relationships, the main goal would have been to find a mate with the most promising genes and the very best survival skills. Personal compatibility would have been welcome but secondary.

These old forces still drive a lot of our behavior. Most evolutionary psychologists, for example, hold that prehistoric women, their mobility hampered by pregnancy and mothering, would have gravitated to men who were strong hunters, high in the tribal pecking order, and likely to be the sort of fathers upon whom they could rely to help parent their offspring. If the realities of childbearing made them more dependent, then it would have made sense for females to gravitate to power and fidelity, probably in that order. A strong mate—physically, socially, and mentally—not only carried promising genes, but also enough strength and social power to put a couple in a position to provide well for their children. We see the resonances of this sort of decision-making everywhere today. There are few women who would avoid a man who is honest, reliable, powerful, and wealthy.

Prehistoric men, on the other hand, would have preferred women who were healthy and who housed nice strings of DNA. They would have looked for clear signs of physical beauty—features like round bottoms, ample breasts, shapely hips, a winning smile—all clear signals of health, especially in a brutal world that battered even the strong. Not much has changed here.

These traits may help explain some of the differences scientists see in the brains of men and women. One brain scanning study conducted recently by scientists at the University of California at Irvine and the University of New Mexico reveals that while men and women are of equal intelligence, they have significantly different brain anatomies. Men have about six and a half times the gray matter related to general intelligence than women have, but women have almost ten times the amount of white matter related to intelligence than men. Gray matter generally provides the neuronal wattage for the brain's information processing centers. White matter provides the synaptic connections between these processing centers.

The team that conducted the study speculates that this helps explain why "men tend to excel in tasks requiring more local processing—like mathematics—while women tend to excel at integrating and assimilating information from distributed gray-matter regions in the brain, such as required for language facility." Very possibly these differences evolved because our male and female ancestors dealt with different environmental and social pressures.[16]

The conclusion of the paper was that evolution seems to have come up with different brain anatomies in each sex that get the same job done just as well, but in different ways. This may also mean that men and women experience the world and one another in fundamentally different ways.

The more interconnected nature of women's brains and their greater facility with language, for example, may explain why women often seem better at communicating what is on their minds, and at the same time they seem to have a knack for reading what is on the minds of others. Cambridge University psychologist Simon Baron-Cohen has found, for example, that when men look at the world, they start with the very small and work outward. Women, on the other hand, tend to get the big picture first and then work inward. Do these views simply reflect the different structures of our brains—men's localized and focused, women's distributed and highly interconnected?

Baron-Cohen, who heads Cambridge's Autism Research Centre, says the research shows that the male brain acts more like an autistic's than a woman's does. Several studies reveal a correlation between the localized, less interconnected nature of men's brains, and their ability to excel at mathematical reasoning and tests that require the mental rotation of three-dimensional objects.[17] Men, says Baron-Cohen, focus first on minute detail, and operate best with a certain detachment. They like to systemize and organize objects, and they tend to be less people-oriented.

This behavior resembles autistics, who are "mindblind," clueless as to the feelings of other people, as if their mirror neurons are malfunctioning. Like men, they also love to systematize and organize, except far more aggressively. Most autistics are also male—overwhelmingly so, in fact. Of those who suffer from Asperger's syndrome, a milder form of autism, there are ten male victims for every female.[18]

These peculiarly male cerebral attributes also may explain why men are more verbally impatient than women (they tend to give orders rather than negotiate), or why they navigate according to an imagined geometric grid while women mostly personalize space by finding familiar landmarks. And it may explain why boys like to play with trucks and guns as opposed to dolls with which they sit down to tea for imagined conversation. They have more of an affinity for things than for people.

But dolls and tea play precisely to women's perceptual and mental skills, which are less analytical, more intuitive, and top-down rather than bottom-up. The interconnected nature of women's brains seems to help them more intuitively read other people. Females are more gifted at absorbing the context of situations, inferring intentions, and then responding in emotionally appropriate ways.

The brain's ability to empathize and socialize is far from well understood, but scientists have found that a cluster of neurons dedicated to reading the feelings and intentions of others lies on the left side of the brain, near Broca and Wernicke's areas, cerebral areas that are usually more developed in women.[19] Altogether this seems to make the females among us more adept social animals, hardwired to get the big picture.

Given what we have gleaned over the years about prehistoric human life, the divergence of male and female brains along these lines makes sense. Men would have done the bulk of the hunting (studies even reveal that men have a very specialized talent that women do not generally have for hitting targets when they throw something at it); they also would have tended to be more analytical, direct, and focused, and less social—all traits that would have favored their survival in a dangerous environment as they foraged and hunted for food. And they would have tended to do the hunting because they were not the sex that became pregnant. They were simply more mobile more of the time.[20]

The reality that women and not men get pregnant would have shaped other roles and behaviors, too, ones in which we often remain entangled today. Women have a limited number of eggs, for example, and pregnancy was (and remains) costly on every possible level. As a result, our female ancestors had to be very careful about the mates they chose. The high stakes would have made the competition among women more fierce. But under the circumstances it could rarely become overt because women in camp also would have needed to get along. A delicate balance had to have been struck in a world where your friends were also your potential competitors. Thoughtfulness and adept social communication would have favored the survival of the women who possessed those traits, both with their children and then with one another. (Maybe this is why mothers, as a group, tend to be more nurturing than fathers.)

While men were dealing with the vicissitudes of the hunt and the dangers that go with it, females would have been faced with handling the complex social world at camp—getting along with one another and handling the kids at the same time they were keeping an eye out for competitors and positioning themselves adeptly in the social hierarchy. If a female was adroit at tuning into her role among those around her, if she had a knack for knowing who her friends and enemies were and a talent for building useful alliances, she would have tended to do quite well.[21,22]

For men the situation was different. Not radically, but different enough. They also would have needed to cooperate with one another on a day-to-day basis while simultaneously competing for the best possible sexual partners.

But since men have millions of sperm and evolved to spread their DNA wherever they could find a willing partner, *and* since they didn't pay the heavy price of pregnancy, they might not have tended to be quite as careful about the females they mated with. (Men generally remain less careful about this today, for the simple reason that if sex leads to pregnancy, the man is not the one who becomes pregnant.)

In this sense, men's choices were less complicated. It was not as important for them to find females who were "true." The main thing was to find females, period, and the healthier the better. This may explain why men seem to be more preoccupied with beauty, shape, and external appearance than women. They were the best indicators of physical health, and to men, physical health was what mattered most—find the best set of complementary chromosomes.

The point is, if form follows function, then differing circumstances may well have shaped different brains, and in one arcing feedback loop those different brains have in turn generated different behaviors and attitudes toward sex and relationships that continue to make the interactions of men and women both so complementary and so complicated. Yet the primal need to bring more children into the world requires that we find a way to meld minds and hearts as well as bodies, even if neither sex sees the world in quite the same way.

. . .

Recently evolved differences in our cerebral anatomy are not the only influences that affect whom we decide to kiss. The limbic system, which is another ancient part of our brains that is well connected to more recently evolved areas, also has a lot to say about why we react to one another the way we do. The limbic system generates a lot of what we label emotion and feeling. When we miss someone, feel warm in the company of a lover, grow angry, jealous, or overjoyed—these experiences are generated by the chemical and electrically charged neuronal conversations that incessantly rattle around our limbic systems.

The whole brain is massively interconnected, but the limbic system seems to be a particularly well-traveled crossroads, a place where the very ancient parts of our brain that govern extremely primal operations like breathing, pleasure, fear, and hunger meet with the high-end thinking and planning we do in the prefrontal cortex. Because of our limbic system it is difficult to disconnect emotion from our memories, actions, or decisions. It also explains why intense emotions reveal themselves in physical reactions like a fast-beating heart, trembling hands, dilated pupils, perspiration, sudden waves of nausea, or equally sudden bursts of joy.

The limbic parts of us help explain a lot of apparently senseless human behavior. Presidents, premiers, senators, and royalty have famously let their limbic systems get the better of them. Think of the headlines that announced the trysts of Senator Gary Hart, President Bill Clinton, or Louisiana Representative Robert Livingston. Even when the career stakes are as high as they can get, this part of the brain has a way of making sure it is heard.

IQ apparently has no bearing on the dictates of our primal drives.

Because the limbic system governs emotional memory, it is a crossroads in another way. It links adult experiences such as kissing and pheromones, sex and love to the memories we gather and carry with us from our childhoods. We have a sense of time and self partly because of our limbically charged memories. They guarantee that many of the emotional patterns from our past, even our very early past, are emotionally saved and colored for the present and the future.

Maybe this is why it is so difficult to write sensibly and logically about kissing. It sits at the nexus of high intellect and primal drives, and the two can't seem to quite get into sync. Kissing symbolizes the entanglements of the human heart. It's an object lesson in the ways lust and love collide in the human race.

Why does this happen to us but not to emus, sea turtles, and the Indri lemurs of Madagascar? Because beginning with the first tools *Homo habilis* fashioned more than two million years ago, human culture has evolved far more rapidly than human DNA. Today, despite all of the genetic rearrangements natural selection has wrought, many of our primal drives remain fully intact. Yet our big brains, in one of the great ironies of the human condition, have entirely refashioned the world into a place profoundly different from the one we originally evolved to live in.

The truth is that kissing, pheromones, and the limbic systems to which they link us often place our feet in each of these camps: the primal one for which we evolved, and the modern one we invented. One shaped by our DNA, the other created by our big brains. Sometimes they seem to be diametrically opposed. It is not that simple, of course, but it is difficult to dispute that the blistering speed of our cultural evolution has placed our DNA and the brains it created at odds.

Genes, for example, demand that we procreate at the earliest possible age. But today sex at age thirteen in many cultures is far from appropriate. Modern teenagers don't have the same career agendas they did 190,000 years ago. But they still have the same bodies and drives.

There are other examples. Partly because of advances in hygiene and medicine, many of us in so-called First World countries can count on living well into our seventies, eighties, and nineties. Monogamous relationships can go on for forty, fifty, or sixty years, far longer than our short-lived ancestors could ever have imagined. As a result, we look not simply for enjoyable sex in a partner that results in new offspring, we also look for fulfilling relationships that can hold up and blossom over the decades. Were we built for that?

Finding and building enduring relationships has not been easy, at least partly because of our limbic and DNA-shaped drives. We value monogamy and fidelity, yet the biggest moneymaker on the Internet is pornography, affairs between both sexes in the modern world are on the rise, and half of all marriages end in divorce in the United States.

A perfect object lesson in the limbic/DNA-driven conflicts that kissing symbolizes is jealousy—the "green-ey'd monster," as Iago so devilishly put it to Othello, "which doth mock the meat it feeds on." Like so many of our primal drives (including kissing itself), jealousy has a hijacker's ways about it. It commandeers parts of the brain like a virus takes control of a cell's genetic machinery. We have even, on occasion, accepted the power of this emotion into our courts of law. In one of the most celebrated cases of the twentieth century, an industrialist from Pittsburgh named Harry K. Thaw pleaded, for the first time in legal history, that he was temporarily insane when he walked up to New York's leading architect, Stanford White, one summer night in 1906 and in full view of a restaurant filled with diners shot him dead at point-blank range.

He did it, he later testified, because a few years earlier White had once enjoyed a relationship with the woman to whom Thaw was married, the great beauty Evelyn Nesbit. A storm erupted in his brain, his lawyers argued at trial, and Thaw had lost all control.[23] The defense worked. Thaw was found not guilty and walked. He was the first, but not the last, person to escape prison on the grounds that he had lost his mind.

When we feel jealous—and almost all of us have felt at least a twinge, if not murderous tides of it—it does seem as though the weather has suddenly changed in our brain. And in some sense it has, if you can describe weather as coursing hormones, activated brain cells, and agitated cerebral molecules.

But it is because the limbic system is connected to our prefrontal cortex that something like jealousy or envy can evolve into premeditated acts of murder or revenge. Using the newer parts of our brain, we can jealously imagine the very worst possible scenarios that in turn feed back to further agitate limbically connected centers such as the hypothalamus and amygdala and hippocampus, which are themselves linked to even more ancient parts of the brain that kick in some of our most primal behaviors—rage, anger, fear. In short, our intellect amplifies the primeval parts of us. A storm in our brain is not a bad analogy.

Truth, Beauty, and the Archaeology of Desire

Who would have thought that the shape of hips and shoulders would so deeply affect the shapes of brains and the evolution of species?

For aeons sexual selection has been tweaking our inborn definitions of beauty and attractiveness, and today those definitions still drive much of our personal behavior as well as explain a lot of what we seem to value in mainstream culture. Cultures everywhere do differ in what they find attractive. Plumper bodies may be more valued in one country than another, and clothing, jewelry, and hairstyles differ in popularity from place to place. But there are some basic elements that all people everywhere, even infants, find attractive. The common determinant seems to be that the traits indicate health of one kind or another, particularly as it relates to the opposite sex, because health is the outward evidence of strong genes, and strong genes lead to individuals who are likely to survive long enough to pass their DNA along to the next generation. In the end, that is the evolutionary bottom line.

Facial symmetry, for example, is universally valued as healthful and therefore attractive. Faces that are symmetrical (proportions between chin and mouth, mouth and eyebrows, etc., that calculate to what the Greeks called the golden number, 1.618 . . .) are considered attractive in all cultures. Often women who have faces that look childlike are also considered attractive—wide eyes and small noses are examples. Another "golden" proportion is a woman's hip ratio. In this case a waist that is 70 percent the size of a woman's hips have become valued because women with those bodies are more likely fertile, and healthy enough to carry a fetus to full term. (All of these calculations are made unconsciously, of course.)

Some studies show that the favored shape of a woman's breast is a three-dimensional parabola rather than a hyperbola, or even a sphere. On the other hand, the preferred shape of a buttocks in a man or a woman is a cardioid, which is the inverse of a parabola.

Long hair in women is often valued because the ability to grow long hair indicates health. The same can be said for fingernails, rosy cheeks, red lips, and clear skin. Accentuating all of these traits are the bedrock of the cosmetics industry, but valued even in countries where there are no advanced cosmetics. On the other hand, jewelry, piercings, and tattoos can also be considered enhancements.

Women have their preferences, too. They almost universally find taller men attractive, especially if they are at least a few inches taller than they are. The theory is that taller men are more dominant, and if they are, they will bring the DNA and power to the union that will help their offspring survive and flourish. For the same reasons, women generally will prefer men whose chests, shoulders, and arms are slightly larger than the average. They might also find beards or other facial hair attractive because they can make a male appear more fierce and dominant (though in the case of Native Americans this wouldn't be true because they can grow virtually no facial hair). An erect posture is valued by both men and women. Erectness is a signal of health and dominance. People slump in defeat or when they aren't feeling well.

In the end, however, physical attractiveness, though powerful, isn't the only arbiter of desire. Physical attributes can be overridden by the personality traits of any particular person, and more than looks make the person. If they are confident, pleasant, and loaded with charisma, pure physical attractiveness becomes secondary.

Given the differing situations of our male and female ancestors, and the differing brain anatomies we have evolved, some evolutionary psychologists theorize that the triggers for jealousy in men and women have developed along different lines. David Buss at the University of Texas (among others) believes that a specific set of brain circuits evolved in men that makes them innately predisposed to jealousy over a mate's *sexual* infidelity. In women, however, researchers have theorized that different circuits trigger jealousy when their mates are *emotionally* untrue.[24]

Christine Harris, a psychologist at the University of California at San Diego, suspects there is more to jealousy than that. She has studied the motives behind murder among couples (5,225, to be exact) across 20 different cultures, and found that there was no real difference behind why men or women did away with their lovers. And in another study, she found that both men and women said the emotional aspects of cheating were more upsetting than the sexual ones. In other words, it wasn't just that their lover had copulated with someone else. Both sexes were driven to murder because they couldn't stomach the idea that someone they loved may have loved someone else. This is Othello in a nutshell.

To find the evidence that jealousy is hardwired into us, we don't have to go back any further than our own infancy. Anyone who has a brother or a sister knows this. Study after study has shown that sibling rivalry is ubiquitous. One study, at Texas Tech University, revealed that infants as young as six months did not like it even one little bit when they noticed their mothers paying more attention to a lifelike baby doll than to them. They furled their eyebrows, fidgeted, turned their lips down, and generally had their limbic systems working overtime to send messages to Mom that they were not happy. And this is among babies who didn't even *have* a brother or a sister. A second study revealed that eight-month-olds will verbally and physically do whatever they can to distract their mothers so they will stop interacting with another child—whine, cry, laugh, whatever works.[25,26]

It's easy to see the roots of adult jealousy and envy in these early reactions, but they didn't evolve because dysfunctional emotional responses do us much good. They evolved because they are survival techniques, and hardwired ones at that, considering the ages of the infants in the studies.

How is this a survival technique? If you are among the most helpless of living mammals and you notice your primary source of safety is not paying attention to you, getting Mom's attention suddenly becomes extremely important. Babies in the past who could keep their mother's focus on them would have survived more often than infants who couldn't, and those genes would have been passed along. Adult jealousy simply demonstrates that we took this technique and found new ways to apply it, mostly dysfunctionally, to our adult relationships. That it sometimes results in murder only illustrates how strong the drive is.

These patterns, etched in our limbic systems during our childhood, have broader effects, too, according to psychiatrists Thomas Lewis, Fari Amini, and Richard Lannon. The concentrated knowledge of our youth "whispers to a child from beneath the veil of consciousness," they write, "telling him what relationships are, how they function, what to anticipate, how to conduct them." Later we apply, for better or worse, these past lessons unconsciously to our current relationships. If this "limbic patterning" gets too far off kilter, we can be in for a lot of personal suffering: "boy meets girl, who (reminiscent of his mother) is needy and stifles his independence; they struggle bitterly over the years and resent each other a little more every day."[27]

If this view of human interaction is accurate, it means that our limbic systems know more about what we want than our conscious minds do. It means that escaping the gravitational forces our caregivers and loved ones created in our youth is tough, lifelong work, and the imprints can never be totally eradicated. Sometimes this might be good, sometimes not. Who can say what kind of leader, husband, and father Winston Churchill would have been had he actually been raised by his politician (later insane) father and socialite mother, both of whom kept their distance during his boyhood?

Perhaps it was best that his beloved nanny, Elizabeth Anne Everest, led him from infancy into adolescence.

We can modify the experiences we had as children, of course. Unlike crickets and frogs, we are not governed solely by our DNA, or even our earliest, most powerful influences. We are consummate learners, after all, and over time we can change the way we behave to get our lives and loves right. In many ways maturing is about controlling and modifying our primal drives so that we can draw power from them instead of being done in by them.

It would be an altogether different and considerably less violent world if the limbic systems of every child emerged into adulthood untrammeled. On the other hand, Shakespeare, Jane Austen, Tolstoy, Hemingway, Woody Allen, and Alfred Hitchcock all would have been robbed of the fascinating, conflicted, obsessed characters they created to keep us in our seats and turning the pages of their unforgettable works. All of literature and entertainment has been built on the backs of our limbic systems and the conflicts they create.

. . .

With our bulging cerebral cortex developed and folded like an old mitt around the more ancient baseball of our limbic brains, we are in some ways back to Plato's two horses: reason and passion. Pheromones, hormones, and dopamine, the nerve endings of lips and tongues, and the pleasure centers they activate speak exquisitely to the primal, emotional parts of us—the parts over which we have little conscious control because their workings mostly hum beneath the radar of our forebrain. Yet the prefrontal cortex is there with its higher centers trying to assess, curtail, manage, and negotiate the older drives.

The combination of the two has given us our greatest art and our most heinous crimes; peace and war; our finest moments and our most deplorable. Without both, enmeshed as they are in us, serial murderers, Hitler, and the architects of the Inquisition would never have been able to justify executing the perfectly innocent people they did. Their actions required both rage and elaborate rationalization. But it is also true that Beethoven would never have conceived and written something as soaringly beautiful as his Ninth Symphony, nor Bach created his Toccata and Fugue in D Minor if our brains were incapable of melding mind and heart, emotion and intellect.

This may be the great gift that kissing gives us. It might have started as a way to share pheromones and help us find the most physically complementary mate. Perhaps it still serves that purpose. But a kiss also fuses both love and passion into one fully, and uniquely, human experience; it bonds people in ways no other human act can. It can open the door to love, the finest of human experiences. It may be true that time and again throughout our lives we will continually pull out the limbic blueprint we were handed as children, and unconsciously match our adult loves to it, but thankfully our intellects also make us great learners capable of remarkable change.

Is this why, even in today's antiseptic world, a kiss sometimes seems so crazy and uncontrollable, so primeval, yet so warm and safe and loving? In one soulful meeting of our lips we can capture all of the colliding forces that shape the core of the human condition, and our personal lives—heart and mind, DNA and intellect, lust and love. This is the limbic system at work—primal, uncontrollable, emotional—run by a brew of pheromones that light up both the brain and the heart. Drawn this close together, our ancient chemical cocktails take the wheel and leave intellect in the dust. Maybe this explains why we sometimes seem to lose our minds when we are entangled with the one we love, and why at those moments we revel and bathe in the insanity of it. And to hell with anything that makes sense.

. . .

What strangely amalgamated creatures we are. What mystifying pieces of evolutionary work. It's odd to think that any creature's future can be directed by such seemingly simple things as knobby or nimble appendages. Or that tears should reveal so much about the complexities of the human heart. From the outside they would hardly seem worth a second glance. But that is the way evolution works. The random scramblings of DNA, shifting climates, jungles in retreat, even mountains that moved, all led to the big toe that allowed our primate ancestors to stand. And that adaptation, in turn, changed how our ancestors related socially and sexually, revamped the way we were born, and created a new kind of primate brain. The same big toe made thumbs—and the tools they fashioned—possible, which led to the evolution of minds capable of language, arguably the greatest tool of all. After all, language enabled us to harness many minds together and create culture; at the same time it transformed us into the self-conscious species, beings with minds resolutely aware of ourselves as well as the world all around.

Yet there is more to us than the logic and sense of language, and the technologies forged by our toolmaking hands and brains. We are built on the genetic foundations of wild animals, and many of our primal drives remain with us right down to the core of our being. They are the wellsprings of the passions, fears, and needs that make us creative, complex, and socially bound. Words, remarkable as their emergence has been, simply are inadequate to express many of our deepest feelings. This is why we not only speak to one another, but kiss and cry and laugh, and dance and paint and make music.

In the end it is difficult to make sense of how we came to be who we are, yet we seem bound and determined to figure it out. Maybe we will never get there. Maybe it doesn't matter. Maybe the thing we love the most is the hunt. And it is that drive that makes us apply our sharp minds to bringing mysteries to heel. If so, it seems we need both for the job—animal passion and human intellect, ancient strands of DNA combined with newly shifted versions to understand the remarkably odd and oddly remarkable beings we are. Human beings.

Epilogue

Cyber sapiens:
The Human Race, Version 2.0

Today nature has slipped, perhaps finally, beyond our field of vision.
—O. B. Hardison Jr.

So now, after six million years of evolution, where do we go next? How will evolution, our newly arrived intellect, our primal drives, and the powerful technologies we continually create, change us in the future?

Our current situation is unlike anything nature has seen before because we are not simply a by-product of evolution, we are ourselves now an agent of evolution. We are this animal, filled with ancient emotions and needs, amplified by our intellects and a conscious mind, embarking on a new century where we are creating fresh tools and technologies so rapidly that we are struggling to keep pace with the very changes we are bringing to the table.

Where will this lead? Will we develop still newer clusters of neurons, new appendages, revamped capabilities just as we have over the past six million years? Absolutely, but probably not in the way you might suspect. It appears, if we look closely, that the DNA that has been such a perfect ally to the changes evolution has brought to us, may itself be in for a revamping. Evolution may be prowling for a new partner. And the partner may be us, or at least the technologies we make possible.

The irony is that evolution requires a being like us, a human being, to bring about changes this fundamental. The job requires an amalgamation of high intelligence and emotion, conscious intent, primal drives, and great quantities of knowledge made possible by minds that can communicate in highly complex ways. If you pulled any one of these out, the future, at least one involving intelligent, conscious creatures like us, would fall apart. It takes not just cleverness, but passion, sometimes fear, fired by focused intention to create and invent. Without this combination there would be no technologies, no wheels or steam engines or nuclear bombs or computers. And there would be nothing like the world we live in today. At best we would still be huddled in the black African night, eking out whatever existence the predators waiting in the darkness around us would allow. Not even fire would be our friend.

But the traits that have shaped us into the human beings we are have endowed us with strange abilities, and they are hurtling us into a future radically unlike the past out of which we have emerged. That future will be profoundly different from anything most of us can imagine.

Take the thinking of Hans Moravec as an object lesson. Moravec is a highly respected robotics scientist at Carnegie Mellon University. In the late 1980s, he quietly passed his spare time writing a book that predicted the end of the human race. The book, titled *Mind Children,* didn't predict that we would destroy ourselves with nuclear weapons or rampant, self-inflicted diseases, or undo the species with self-replicating nanotechnology. Instead, Moravec, who had an abiding and life-long fascination with intelligent machines, predicted that we would invent ourselves out of existence, and robots would be the technology of choice.

In a subsequent book, *Robot: Mere Machine to Transcendent Mind,* Moravec explained that this transformation would unfold one technological generation at a time, and because of the blistering rate of change today, would pretty much run its course by the middle of the twenty-first century. We would manage this by boosting robots up the evolutionary ladder, roughly in decade-long increments, making them smarter, more mobile, more like us. First they would be as intelligent as insects or a simple guppy (we are about there right now), then lab rats, then monkeys and chimps until finally one day the machines would become more adept and adaptive than their makers. That, of course, would quickly raise the question "Now who is in charge?" Would *Homo sapiens,* after some two hundred thousand years of living on top of the planet's food chain, no longer rule the roost? Would we, in the cramped space of this evolutionary ellipsis, find ourselves playing Neanderthal to technologies that had become, like us, self-aware—the first conscious tools built by a conscious toolmaking creature?

The unavoidable answer would be yes. Evolution will have found, through us, a new way to make a new creature, one that could forsake its ladders of DNA and the fragile, carbon-based biology that nature had been using for nearly four million millennia to manage the job.

The "end" would not come in the form of a *Terminator*-style invasion; it would simply unfold in the natural course of evolutionary events where one species, better adapted to its environment, replaces another that is no longer very fit to continue. Except the new species wouldn't be cobbled out of DNA, it would be fashioned from silicon and alloy, invented by us, and once successfully brought into the world, our species would no longer be required.

Whether events will play out like this or not remains to be seen. But Moravec's scenario makes a point: The world and the life upon it changes, and simply because we are the agents of change doesn't mean we won't be affected by it.

.   .   .

It is strange to think of the invention of machines, even robotic ones, as having anything to do with Darwin's natural selection. We usually regard evolution as biological—a world of cells, DNA, and "living" creatures. And we think of our machines as unalive, unintelligent, and shifted by economic forces more than natural ones. But it isn't written anywhere that evolution has to be constrained by what we traditionally think of as biology. In fact, each day the lines between biology and technology, humans and the machines we create, are blurring. We are already part and parcel of our technology.

Since the day *Homo habilis* three-jaw-chucked his first flint knife, it has been difficult to know whether we invented our tools or our tools invented us. The world economy would crash if its computer systems failed. We can't live without laptops, palmtops, cell phones, or iPods, which grow continually smaller and more powerful. We regularly engineer genes, despite the raging debates over stem cell therapy. A human being will very likely be cloned within the next five years. We now have computer processors working at the nano (molecular) level and microelectromechanical machines (MEMS) that operate at cellular dimensions. Already electronic prosthetics make direct connections with human nerves, and electronic brain implants for Parkinson's disease and weak hearts are commonplace. Scientists are even experimenting with electronic, implantable eyes. New clothing weaves digital technologies into their fiber and brings them a step closer to being a part of us. The military is working on a "battlesuit," a kind of second skin that will amplify a soldier's senses, strength, and ability to communicate, even triangulate the direction of a bullet headed his or her way.

What next? Speech, writing, and art enabled us to share inner feelings in new and powerful ways. But it takes months or years to learn a new language or how to play the piano or master the art of engineering bridges and buildings. Will new technologies that accelerate communication (virtual reality, telepresence, digital implants, nanotechnology) create new ways to communicate that can bypass speech? Will we someday communicate by a kind of digital telepathy, downloading information, experiences, skills, even emotions the way we download a file from the Internet to our laptop? Will we become machines, or will machines become more powerful versions of us? And if any of this comes to pass, what ethical issues do we face? At what point to do we stop being human?

Lynn Margulis, arguably the world's leading microbiologist, has argued that this blurring of technology and biology isn't really new at all. She has observed that the shells of clams and snails are a kind of technology dressed in biological clothing.* Is there really that much difference between the vast skyscrapers we build or the malls in which we shop, even the cars we drive around, and the hull of a seed? Seeds and clamshells, which are not alive, hold in them a little bit of water and carbon and DNA, ready to replicate when the time is right, yet we don't distinguish them from the life they hold. Why should it be any different with office buildings, hospitals, and space shuttles?

Put another way, *we* may make a distinction between living things and the tools those things happen to create, but nature does not. The processes of evolution simply witness new adaptations and preserve those that perform better than others. That would make *Homo habilis*'s first flint knife a form of biology as sure as a clamshell, one that set our ancestors on a fresh evolutionary path, just as if their DNA had been tweaked to create a new, physical mutation—say, an opposable thumb or a big toe.

Even if these technological adaptations were outside what we might consider normal biological bounds, the effect was just as profound and far

* This was during a conversation with Professor Margulis at her home in western Massachusetts.

more rapid. In an evolutionary snap, that first flint knife changed what we ate and how we interacted with the world and one another. It enhanced our chances of survival. It accelerated our brain growth, which in turn allowed us to create still more tools, which led to yet bigger brains. And on we went, continually and with increasing speed and sophistication, fashioning progressively more complex technologies right up to the genetic techniques that enable us to fiddle with the selfsame ribbons of our chromosomes that made the brains that conceived tools in the first place. If this is true, then all of our technologies are an extension of us, and each human invention is really another expression of biological evolution.

Moravec and Margulis aren't alone in asking questions that force us to bend our traditional thinking about evolution. Scientist and inventor Ray Kurzweil has, like Moravec, pointed out that the rate of technological change is increasing at an exponential rate. Also like Moravec, he foresees machines as intelligent as we are evolving by midcentury. Unlike Moravec, he doesn't necessarily believe they will arrive in the form of robots.

Initially Kurzweil sees us reengineering ourselves genetically so that we will live longer and healthier lives than the DNA we were born with might normally allow. We will first rejigger genes to reduce disease, grow replacement organs, and generally postpone many of the ravages of old age. This, he says, will get us to a time late in the 2020s when we can create molecule-sized nanomachines that we will program to tackle jobs our DNA never evolved naturally to undertake.

Once these advances are in place we will not simply slow aging, but reverse it, cleaning up and rebuilding our bodies molecule by molecule. We will also use them to amplify our intelligence, nestling them among the billions of neurons that already exist inside our brains. Our memories will improve; we will create entirely new, virtual experiences on command, and take human imagination to levels our currently unenhanced brains can't begin to conceive. In time (but pretty quickly), we will develop into a completely digital species that has reverse-engineered the human brain into a vastly more powerful, digital version.

This view of the future isn't fundamentally different from Moravec's brain-to-robot download, except it is more gradual. Either way we will have melded with our technology if, in fact, those barriers ever really existed in the first place, and in the end, erase the lines between bits, bytes, neurons, and atoms.

Or looked at another way, we will have evolved into another species. We will no longer be *Homo sapiens,* but *Cyber sapiens*—a creature part digital and part biological that will have placed more distance between its DNA and its destiny than any other creature. And we will have become a creature capable of steering its own evolution ("cyber" derives from the Greek word for a ship's steersman or navigator—*kybernetes*), an entirely new state of af- fairs in the natural world.

Why would we allow ourselves to be displaced? Because in the end, we won't really have a choice. Our own inventiveness has already unhinged our environment so thoroughly that we are struggling to keep up. In a supreme irony we have created a world fundamentally different from the one into which we originally emerged. A planet with six and a half billion creatures on it, traveling in flying machines every day by the millions, their minds roped together by satellites and fiber-optic cable, rearranging molecules on the one hand and leveling continents of rain forest on the other, growing food and shipping it overnight by the trillions of tons—all of this is a far cry from the hunter-gatherer, nomadic life for which evolution had fashioned us two hundred thousand years ago.

So it seems the long habit of our inventiveness has placed us in a pickle. In the one-upsmanship of evolution, our tools have rendered the world more complex, and that complexity requires the invention of still more complex tools to help us keep it all under control. Our new tools enable us to adapt more rapidly, but one advance begs the creation of another, and each increasingly powerful suite of inventions shifts the world around us so powerfully that still more adaptation is required.

The only way to survive is to move faster, get smarter, change with the changes; and the best way to do that is to amplify ourselves eventually right out of our own DNA so we can survive the new environments—physical, emotional, and mental—that we keep creating.

Is all of this too implausible to consider? Will *Homo sapiens* really give way to *Cyber sapiens* who seamlessly integrate the molecular and digital worlds just as our ancestors merged the technological and biological worlds two million years ago? Evolution has presided over stranger things. It took billions of years before the switching and swapping of genes brought us into existence. Our particular brain then took two hundred thousand years to get us from running around in skins with stone weapons to the world we live in today. Evolution is all about the implausible. And the drive to survive is a relentless shaper of the seemingly impossible. We ourselves are the best proof.

If all of this should happen; if DNA itself goes the way of the dinosaur, what sort of creature will *Cyber sapiens* be? In some ways we can't know the answer any more than *Homo erectus* could imagine how his successors would someday create movies, invent computers, and write symphonies. Our progeny will certainly be more intelligent, with brains that are both massively parallel, like the current version we have, and unimaginably fast. But what of those primal drives that we carry inside our skulls, and those nonverbal, unconscious ways of communicating? What of laughter and crying and kissing? Will *Cyber sapiens* know a good joke when he hears one, or smile appreciatively at a fine line of poetry? Will he tousle the machine-made hair of his offspring, hold the hand of the one he loves, kiss soulfully, wantonly, and uncontrollably? Will there be a difference between the "brains" and behaviors of he and she? Will there even *be* a he and a she? And what of pheromones and body language and nervous giggles? Maybe they will have served their purpose and gone away. Will *Cyber sapiens* sleep, and if they do, will they dream? Will they connive and gossip, grow mad with jealousy, plot and murder? Will they carry with them a deep, if machine-made, unconscious that is the dark matter of the human mind, or will all of those primeval secrets be revealed in the bright light cast by their newly minted minds?

We may face these questions sooner than we imagine. The future gathers speed every day.

I'd like to think the evolutionary innovations and legacies that have combined to make us so remarkable, and so human, won't be left entirely behind as we march ahead. Perhaps they can't be. After all, evolution does have a way of working with what is already there, and even after six million years of wrenching change, we still carry with us the echoes of our animal ancestors. Maybe the best of those echoes will remain. After all, as heavy as some baggage can be, preserving a few select pieces might be a good thing, even if we are freaks of nature.

Acknowledgments

As solitary an exercise as writing appears to be, it is never really done alone. The making of books requires a kind of community effort, and this one was no exception. So my sincere thanks go out to those who helped make possible the pages that you have read. The scientists, for example, who did so much of the heavy intellectual lifting in so many fields, from the fascinating work of Terrence Deacon, Michael Arbib, and Giacomo Rizzolatto on the evolution of language to the startling insights into human consciousness that Michael Gazzinga, Oliver Sacks, and Gerald Edelman have developed. There is the work of Donald Johanson, Ian Tattersal, the Leakey family, and the battalions of anthropologists who have sifted through the dust of Asia, Africa, and Europe to uncover clues about our past that help explain how our species managed to make its way into the present. I am grateful for the research into social behavior, laughter, crying, and brain development by Robin Dunbar, Jane Goodall, Randolph Cornelius, Patricia Greenfield, Robert Provine, Henry Plotkin, Dean Falk, and others that helped flesh out areas of our evolution and behavior that a couple of decades ago were hardly understood at all. And generally I am grateful for the work of Steven Pinker, Lynn Margulis, Ray Kurzweil, Hans Moravec, Lewis Thomas, and Richard Dawkins, who have a habit of seeing nearly everything in fresh and inspiring ways. These are only a few of the hundreds of scientists whose work shed light on how we became the creatures we are.

I am also thankful to my agent Peter Sawyer for introducing me to one of the finest people I have ever met, Walker and Company's publisher, George Gibson. The support that both of them have offered on this project has been unwavering and positive. Jacqueline Johnson took on the task of editing my writing. Her endless patience and gentility made that job a pleasure rather than drudgery, and her skill took jumbled passages more than once and made them smooth and coherent.

Several close friends showed just how close they were by reading through the manuscript in its various phases to provide me feedback. I am especially grateful to Richard Tobin, Cyndy Mosites, my father and mother, Bill and Rosemary Walter, Robin Wertheimer, Mary Murrin, Jerry Farber, and Tara McLamey.

Above all I am grateful to my two extraordinary daughters, Molly and Hannah, who so often put up with a father who attended softball games, crew races, and theatrical performances with his mind sometimes still in ancient Africa or mulling over brain anatomy. But they were always patient with me, kept their sense of humor (and mine) tuned up, and were constant reminders of how lucky we all are to be the only creatures that can laugh, kiss, and cry with the ones we love.

—C. W.

Pittsburgh, Pennsylvania

2006

Bibliography

Ackerman, Diane. *A Natural History of the Senses*. New York: Vintage Books, 1990.

Arbib, Michael A. "From Monkey-like Action Recognition to Human Language: An Evolutionary Framework for Neurolinguistics." *Behavioral and Brain Sciences* 28(2) (2005): 105–24.

Berck, Judith. "Before Baby Talk, Signs and Signals." *New York Times*, January 6, 2004.

Bronowski, J. *The Ascent of Man*. Boston: Little, Brown, 1973.

Cane, Mark A., and Peter Molnar. "Closing of the Indonesian Seaway as a Precursor to East African Aridification around 3–4 Million Years Ago." *Nature* 411 (2001): 157–62.

Carey, Benedict. "Can Brain Scans See Depression?" *New York Times*, October 18, 2005.

Chalmers, David J. "The Puzzle of Conscious Experience." *Scientific American* 12(1) (2002): 90–100.

Child's Play. PBS, May 29, 1997.

Coghlan, Andy. "Laughing Helps Arteries and Boosts Blood Flow." *New Scientist* (March 7, 2005).

Conway, William. "The Congo Connection." *Wildlife Conservation* (August 1999): 18–25.

Coqueugniot, H., J.-J. Hublin, F. Veillon, F. Houet, and T. Jacob. "Early Brain Growth in *Homo erectus* and Implications for Cognitive Ability." *Nature* 431 (2004).

Corballis, Michael C. *From Hand to Mouth: The Origins of Language*. Princeton, N.J.: Princeton University Press, 2003.

Damasio, Antonio R. "How the Brain Creates the Mind." *Scientific American* 12(1) (2002): 4–9.

Darwin, Charles. *The Descent of Man*. Norwalk, Conn.: Heritage Press, 1972.

———. *Origin of Species*. New York: Gramercy, 1995.

Dawkins, Richard. *The Blind Watchmaker*. New York: W. W. Norton, 1987.

Deacon, Terrence W. *The Symbolic Species*. New York: W. W. Norton, 1998.

Dennett, Daniel C. *Consciousness Explained*. Boston: Little, Brown, 1991.

———. *Freedom Evolves*. New York: Viking, 2003.

Diamond, Jared. *Guns, Germs, and Steel*. New York: W. W. Norton, 1997.

Dick, Philip K. *Do Androids Dream of Electric Sheep?* New York: Random House, 1996.

"Don't Knock Climate Change—It May Be the Reason We're Here." *Geology News* 23 (January 2001).

Donald, Merlin. *Origins of the Modern Mind—Three Stages in the Evolution of Culture and Cognition.* Cambridge, Mass.: Harvard University Press, 1991.

Downey, Charles. "Toxic Tears: How Crying Keeps You Healthy." *Health Info.* 2000, EBSCO Publishing, Ipswich, Mass.

Dunbar, Robin. *Grooming, Gossip, and the Evolution of Language.* Cambridge, Mass.: Harvard University Press, 1996.

Dyson, Freeman. *Disturbing the Universe.* New York: Harper & Row, 1979.

"Earliest Human Ancestors Discovered in Ethiopia: Discovery of Bones and Teeth Date Fossils Back More Than 5.2 Million Years." *ScienceDaily*, July 12, 2001.

Edelman, Gerald M. *Wider than the Sky: The Phenomenal Gift of Consciousness.* New Haven, Conn.: Yale University Press, 2004.

Edelman, Gerald M., and Giulio Tononi. *Consciousness: How Matter Becomes Imagination.* New York: Penguin Books, 2000.

———. *Universe of Consciousness: How Matter Becomes Imagination.* New York: Basic Books, 2001.

Eiseley, Loren. *The Unexpected Universe.* Orlando, Fla.: Harcourt, Brace, 1969.

Flatt, Adrian E. "Our Thumbs." *Baylor University Medical Center Proceedings* 15 (2002): 380–87.

French, Mary Ann. "Grin and Bare It." *Boston Globe Magazine,* September 11, 2000.

Frey, William, and Muriel Langseth. *Crying: The Mystery of Tears.* San Francisco: Harper & Row, 1985.

Gazzaniga, Michael S. *Mind Matters: How Mind and Brain Interact to Create Our Conscious Lives.* Boston: Houghton Mifflin, 1988.

———. *The Social Brain.* New York: Basic Books, 1985.

———. "The Split Brain Revisited." *Scientific American* 12(1) (2002): 26–31.

Gladwell, Malcolm. *The Tipping Point: How Little Things Can Make a Big Difference.* Boston: Little, Brown and Co. 2000.

Goodall, Jane. *In the Shadow of Man.* Boston: Houghton Mifflin, 1988.

Gould, Stephen J. *Ontogeny and Phylogeny.* Cambridge, Mass.: Harvard University Press, Belknap Press, 1997.

Greenfield, Patricia M. "Language, Tools, and Brain: The Ontogeny and Phylogeny of Hierarchically Organized Sequential Behavior." *Behavioral and Brain Sciences* 14(4) (1991): 531–51.

Grutzendler, Jaime, Narayanan Kasthuri, and Wen-Biao Gan. "Long-Term Dendritic Spine Stability in the Adult Cortex." *Nature* 420 (2002): 812–16.

Haier, Richard J., Rex E. Jung, Ronald A. Yeo, Kevin Head, and Michael T. Alkire. "Structural Brain Variation and General Intelligence." *NeuroImage* 23 (2004): 425–33.

Hall, Edward T. *The Silent Language.* Garden City, N.Y.: Doubleday, 1959.

Harrington, Jonathan. "Phonology and the Structure of Language." Class notes: MacQuarie University, Sydney, Australia, online at: http://www.ling.mq.edu.au/speech/phonetics/phonology/structure/structure.pdf.

Harris, Christine R. "The Evolution of Jealousy." *American Scientist* 92 (2004): 62.

Harris, Marvin. *Our Kind*. New York: Harper & Row, 1990.

Herrmann, Christoph S., Angela D. Friederici, Ulrich Oertel, Burkhard Maess, Anja Hahne, and Kai Alter. "The Brain Generates Its Own Sentence Melody: a Gestalt Phenomenon in Speech Perception." *Brain and Language* 85 (2003): 396–401.

Hickok, Gregory, Ursula Bellugi, and Edward S. Klima. "Sign Language in the Brain." *Scientific American* 12(1) (2002): 46–53.

Horgan, John. *The Undiscovered Mind: How the Human Brain Defies Replication, Medication, and Explanation*. New York: Simon & Schuster, 2000.

"Human Brain Evolution Was a 'Special Event'" *Howard Hughes Medical Institute* 29 (December 2004).

"Human Intelligence Determined by Volume and Location of Gray Matter Tissue in Brain." *UCI Communications*, July 19, 2004.

Hurford, James R. "Language Beyond Our Grasp: What Mirror Neurons Can, and Cannot, Do for Language Evolution," in *Evolution of Communication Systems: A Comparative Approach,* ed. D. Kimbrough Oller and Ulrike Griebel (Cambridge, Mass.: The MIT Press, 2004): 297–313.

Jaynes, Julian. *The Origin of Consciousness in the Breakdown of the Bicameral Mind*. Boston: Houghton Mifflin, 1982.

Johanson, Donald C., and Edey A. Maitland. *Lucy: The Beginnings of Humankind*. London: Penguin, 1981.

Johanson, Donald C., and James Shreeve. *Lucy's Child: The Discovery of a Human Ancestor*. New York: William Morrow, 1989.

Johnson, D. R. "Retardation and Neotony in Human Evolution." *Human Evolution* (November 24, 2004). http://www.leeds.ac.uk/chb/lectures/anthl_06.html.

Johnson, Steven. *Emergence: The Connected Lives of Ants, Brains, Cities, and Software*. New York: Touchstone, 2001.

Kalin, Ned H. "The Neurobiology of Fear." *Scientific American* 12(1) (2002): 72–81.

Kempermann, Gerd, and Fred H. Gage. "New Nerve Cells for the Adult Brain." *Scientific American* 12(1) (2002): 38–44.

Kidd, Robert. "Evolution of the Rearfoot: A Model of Adaptation with Evidence from the Fossil Record." *Journal of the American Podiatric Medical Association* 89 (1999): 2–17.

Kimura, Doreen. "Sex Differences in the Brain." *Scientific American* 12(1) (2002): 32–37.

Knight, Will. "Computer Crack Funnier than Many Human Jokes." *NewScientist* 20 (December 2001).

Kobayashi, Yoshihiro. "Pheromones: The Smell of Beauty." Ph.D. diss., University of Colorado, 1997.

Koontz, Dean. *The Face*. New York: Bantam Doubleday Dell, 2004.

Kurzweil, Ray. *The Age of Spiritual Machines*. New York: Penguin, 1999.

———. *Are We Spiritual Machines?* Seattle, Wash.: Discovery Institute, 2002.

———. *The Singularity in Near*. New York: Penguin, 2005.

Laitman, Jeffrey T. "How Humans and Their Closest Kin Perceive the World: The Special Senses of Primates." *The Anatomical Record* 281A (2004).

Lakoff, George, and Mark Johnson. *Philosophy in the Flesh: The Embodied Mind and Its Challenge to Western Thought*. New York: Perseus, 1998.

Lambert, David, and the Diagram Group. *The Field Guide to Early Man*. New York: Facts On File, 1987.

Leakey, L. S. B., P. V. Tobias, and J. R. Napier. "A New Species of the Genus *Homo* from Olduvai Gorge." *Nature* 202 (1964): 7–9.

Leakey, Meave, and Alan Walker. "Early Hominid Fossils from Africa." *Scientific American* 13(2) (2003): 14–19.

Leakey, R. E. F. "Evidence for an Advanced Plio-Pleistocene Hominid from East Rudolf, Kenya." *Nature* 242 (1973): 447–50.

Ledoux, Joseph E. "Emotion, Memory, and the Brain." *Scientific American* 12(1) (2002): 62–71.

————. *The Emotional Brain*. New York: Simon & Schuster, 1998.

————. *Synaptic Self: How Our Brains Become Who We Are*. New York: Penguin, 2003.

Leonard, William R. "Food for Thought." *Scientific American* 13(2) (2003): 62–71.

Leonard, William R., Marcia L. Robertson, J. J. Snodgrass, and Christopher W. Kuzawa. "Metabolic Correlates of Hominid Brain Evolution." *CBP* 136 (2003): 5–15.

Lieberman, D. E., R. C. McCarthy, K. M. Hiiemae, and J. B. Palmer. "Ontogeny of Post-natal Hyoid and Larynx Descent in Humans." *Archives of Oral Biology* 46 (2001): 117–28.

Lumsden, Charles J., and Edward O. Wilson. *Genes, Mind, and Culture: The Coevolutionary Process*. 25th ed. New York: World Scientific, 2005.

Lutz, Tom. *Crying: The Natural and Cultural History of Tears*. New York: W. W. Norton, 1999.

Margulis, Lynn. *Early Life*. Boston: Jones & Bartlett, 1984.

Margulis, Lynn, and Dorion Sagan. *Microcosmos*. New York: Simon & Schuster, 1991.

Marks, Jonathan. *What It Means to Be 98% Chimpanzee: Apes, People, and Their Genes*. Berkeley: University of California Press, 2003.

Marrin, Minette. "Why Does It All End in Tears?" *Sunday Telegraph*, London, December 16, 2001.

McNeill, Daniel. *The Face*. Boston: Little, Brown, 1998.

Moravec, Hans. *Mind Children*. Cambridge, Mass.: Harvard University Press, 1988.

————. *Robot: Mere Machine to Transcendent Mind*. New York: Oxford University Press, 1999.

Morris, Desmond. *Intimate Behaviour*. New York: Bantam Books, 1973.

————. *The Naked Ape*. New York: McGraw-Hill, 1967.

"The Oldest *Homo sapiens*: Fossils Push Human Emergence Back to 195,000 Years Ago." *ScienceDaily*, February 28, 2005.

Osborne, Lawrence. "A Linguistic Big Bang." *New York Times*, October 24, 1999.

Parachin, Victor. "Fears About Tears? Why Crying Is Good or You." *Vibrant Life* 11 (November 1992).

Perella, Nicolas J. *Kiss Sacred and Profane*. Berkeley: University of California Press, 1969.

Perlman, David. "Fossils from Ethiopia May Be Earliest Human Ancestor." *National Geographic,* July 21, 2001.

Petito, Laura A., Siobhan Holowka, Lauren E. Sergio, and David Ostry. "Language Rhythms in Baby Hand Movements." *Nature* 413 (2001): 35–36.

Petitto, Laura Ann, Robert J. Zatorre, Kristine Gauna, E. J. Nikelski, Deanna Dostle, and Alan C. Evans. "Speechlike Cerebral Activity in Profoundly Deaf People Processing

Signed Languages: Implications for the Neural Basis of Human Language." *PNAS* 97 (2000): 13,961–13,966.

Pinker, Steven. *The Language Instinct.* New York: William Morrow, 1994.

Plotkin, Henry. *Darwin Machines and the Nature of Knowledge.* Cambridge, Mass.: Harvard University Press, 1997.

Provine, Robert. *Laughter: A Scientific Investigation.* New York: The Penguin Group, 2000.

Quartz, Steven R., and Terrence J. Sejnowski. *Liars, Lovers, and Heroes.* New York: Harper-Collins, 2002.

Rayl, A. J. S. "Humor: A Mind-Body Connection." *Scientist,* October 2, 2000.

Reader, Simon Matthew. "Outline and Abstract of 'Brain Size in Primates as a Function of Behavorial Innovation.'" Ph.D. diss., University of Utrecht.

Richter, Jean Paul. *The Notebooks of Leonardo Da Vinci.* Vol. 1. New York: Dover, 1970.

Ridley, Matt. *The Red Queen: Sex and the Evolution of Human Nature.* New York: Harper-Collins, 2003.

Rizzolatti, Giacomo, and Laila Craighero. "The Mirror Neuron System." *Annual Review of Neuroscience* 27 (2004): 169–92.

Rizzolatti, Giacomo, Leonardo Fogassi, and Vittorio Gallese. "Motor and Cognitive Functions of the Ventral Premotor Cortex." *Cognitive Neuroscience* 12: 149–54.

———. "Neurophysiological Mechanisms Underlying the Understanding and Imitation of Action." *Nature* 2 (2001): 661–70.

Rodman, Peter., and McHenry, Henry. "Bioenergetics and Origins of Hominid Bipedalism." *American Journal of Physical Anthropology* 52 (1980): 103–106.

Rosenberg, Karen R., and Wenda R. Trevathan. "The Evolution of Human Birth." *Scientific American* 13(2) (2003): 80–85.

Rossi, William A. "The Foot: Mother of Humanity; Mankind Owes Homage to Our Uniquely Human Feet, Without Which It Could Not Have Evolved to Its Present State." *Podiatry Management,* April 1, 2003.

Sacks, Oliver. *Anthropologist on Mars: Seven Paradoxical Tales.* New York: Random House, 1996.

Sagan, Carl. *The Dragons of Eden.* New York: Ballantine Books, 1977.

Samiei, Haleh V. "Why We Weep." *Washington Post,* January 12, 2000.

Sapolsky, Robert M. *A Primate's Memoir.* New York: Simon & Schuster, 2001.

Schick, Kathy D., and Nicholas Toth. *Making Silent Stones Speak: Human Evolution and the Dawn of Technology.* New York: Touchstone Books, 1993.

Schwartz, Jeffrey H. *Sudden Origins: Fossils, Genes, and the Emergence of Species.* New York: John Wiley & Sons, 1999.

Skoyles, John R., and Dorion Sagan. *Up from Dragons.* New York: McGraw-Hill, 2002.

"Soil Suggests Early Humans Lived in Forests Instead of Grasslands." *ScienceDaily,* July 13, 2001.

Spice, Bryon. "New Species Right out of Box." Pittsburgh *Post-Gazette,* February 21, 1999.

Steinberg, Marlene, and Maxine Schnall. *The Stranger in the Mirror.* New York: Harper-Collins, 2001.

Stout, Martha. *The Myth of Sanity: Divided Consciousness and the Promise of Awareness.* New York: Penguin, 2002.

Tattersal, Ian. *The Fossil Trail: How We Know What We Think We Think About Human Evolution.* New York: Oxford University Press, 1995.

———. "Once We Were Not Alone." *Scientific American* 13(2) (2003): 20–27.

———. "Out of Africa Again . . . and Again?" *Scientific American* 13(2) (2003): 38–45.

Tattersall, Ian, and Jeffrey H. Schwartz. *Extinct Humans.* Boulder, Colo.: Westview, 2001.

Taylor, Timothy. *The Prehistory of Sex.* New York: Bantam Books, 1997.

"Teenagers Special: Brain Storm." *NewScientist* 5 (March 2005).

Thomas, Lewis. *The Medusa and the Snail.* New York: Bantam Books, 1986.

Thorne, Alan G., and Milford H. Wolpoff. "The Multiregional Evolution of Humans." *Scientific American* 13(2) (2003): 46–53.

Trachtenberg, Joshua T., Brian E. Chen, Graham W. Knott, Guoping Feng, Joshua R. Sanes, Egbert Welker, and Karel Svoboda. "Long-Term in Vivo Imaging of Experience-Dependent Synaptic Plasticity in Adult Cortex." *Nature* 420 (2002): 788–94.

Van Hooff, Jan. *The Evolution of Laughter.* Transcript of Lecture: University of Utrecht from a film by Joost de Haas. http://www.uni-duessldorf.de/WWW/MathNat/Ruch/Psy356-Handouts/The%20Evolution%20of%20Laughter.pdf.

Vincent, Jean-Didier. *The Biology of Emotions.* Cambridge, Mass.: Basil Blackwell, 1990.

Wade, Nicholas. "A Biological Dig for the Roots of Language." *New York Times,* March 16, 2004.

———. "Comparing Genomes Shows Split Between Chimps and People." *New York Times,* December 12, 2003.

Weiner, Jonathan. *Time, Love, Memory: A Great Biologist and His Quest for the Origins of Behavior.* New York: Alfred A. Knopf, 1999.

"Why Women Have Breasts." May 6, 2004. http://www.staff.ncl.ac.uk/nikolas.lloyd/evolve/breasts.html.

Wilcox, Sherman. "Gesture, Icon, and Symbol: What Can Signed Languages Tell Us About the Origin of Signs?" Online publication.

———. "The Invention and Ritualization of Language." Albuquerque: University of New Mexico, 1996.

Wills, Christopher. *The Runaway Brain.* New York: HarperCollins, 1993.

Wilson, Edward O. *On Human Nature.* Cambridge, Mass: Harvard University Press, 1978.

Wilson, Frank R. *The Hand.* New York: Pantheon Books, 1998.

Wong, Kate. "An Ancestor to Call Our Own." *Scientific American* 13(2) (2003): 4–13.

Wright, James D. "Climate Change: The Indonesian Valve." *Nature* 411 (2001): 142–43.

Wright, Robert. *The Moral Animal: The New Science of Evolutionary Psychology.* New York: Pantheon Books, 1994.

Young, Emma. "Inside the Brain of an Alcoholic." *New Scientist,* February 4, 2006.

Notes

Prologue

1. This is an estimate from *Microcosmos,* a book written by science journalist Dorion Sagan and his mother, microbiologist Lynn Margulis, and published in 1986.

2. Scientist and inventor Ray Kurzweil has pointed out that as complicated as the human genome is, it contains relatively little information: about three billion rungs or six billion bits or roughly eight hundred millions bytes (with many redundancies). In his book *The Singularity Is Near* he argues that the genome can be compressed to close to thirty million bytes—less than a Microsoft Word program. On the other hand, this fairly simple "program" sets in motion processes that create the human brain, which is a *billion times* more complex than the genome itself. One example is the human cerebellum, which contains nearly half of the brain's neurons, yet only a handful of genes—a few tens of thousands of bytes of information—express the wiring instructions for that part of the brain. Of course, it is the flexibility of this brain and its ability to cull, store, manipulate, and create new information that makes it so remarkable, and made figuring out how to map of the genome was possible in the first place.

3. Carl Sagan, *The Dragons of Eden* (New York: Ballantine Books, 1977), p. 42.

Chapter 1: The Curious Tale of Hallux Magnus

1. Paleoanthropology is not an exact science, and scientists still hotly debate when precisely the single line of apes that we, chimps, and gorillas share actually split into separate evolutionary lines. Most agree that it happened five million to six million years ago, at the close of the Miocene epoch.

2. James D. Wright, "Climate Change: The Indonesian Valve," *Nature* 411 (2001): 142–43.

3. Two recent finds have fueled the debate about the precise timeline and origins of our direct ancestors. One, a skull found in July of 2001 in Chad, fifteen hundred miles west of the Rift Valley, is thought to be close to six million years old. Some say that this fossil, dubbed *Sahelanthropus tchadensis,* is a hominid, the nonape line of creatures from which we evolved. Others say it may be the last common link between chimpanzees and

us. It is difficult to resolve. Only skull fragments were found, which means we have no way of knowing whether the creature walked upright or on all fours. Another creature, *Orrorin tugenensis,* was found in Kenya's Tugen Hills, also in 2001. Its discoverers maintain that *tugenensis* is a hominid. Like *Sahelanthropus tchadensis, Orrorin* has features that are both simian and hominidlike. Based on the information at hand, it's too difficult to say whether they are our direct ancestors. Rick Potts, director of the human origins program at the Museum of Natural History in Washington, D.C., might have summed up the situation best: "A couple of years ago, quite a number of us were simply waiting for *Ardipithicus ramidus* (the oldest undisputed hominid fossil) to tell us what it was all about. We thought that it would be the most primitive hominid. All of this mixing and matching [of traits] suggests a lot of population isolation, independent evolution, and coalescence of populations again. It's going to be really difficult to figure all of this out."

Paleoanthropologists aren't completely in agreement about when our ancestors began to walk and why. The most accepted theory is that hominids began to walk upright after they were forced to deal with the climatic destruction of many of Africa's jungles beginning about six million years ago. At that time, most of Africa was forested. But recent fossil discoveries have called that theory into question. In 2001, anthropologists found fragments of skull, hand, arm, and collarbone fragments from a creature *(Ardipithecus ramidus kadabba)* that lived in the Middle Awash River Valley of Ethiopia almost 5.2 to 5.8 million years ago. Of all the bones they found, though, the most interesting was a single toe bone that indicated this animal could stand upright. What makes this odd is that other fossil studies of the region's soil indicate that six million years ago this part of Ethiopia was forested. So if it was the spread of savannas and grasslands that forced early hominids to stand on their hind legs, why was this creature walking upright when it was still living in a jungle? Maybe the toe is misleading. It may be that beyond the forest where the creature lived there were wide-open spaces that required upright walking. It may mean that this creature was an evolutionary aberration. Some scientists have speculated that early creatures from this period became "preadapted" for walking upright by walking along the branches of large trees. Whatever the case, somehow creatures ancestral to us eventually stood up, evolved in the open savannas of East Africa, and led to the creatures we are today.

4. M. Brunet, et al., "A New Hominid from the Upper Miocene of Chad, Central Africa," *Nature* 418 (2002): 145–51; P. Vignaud, et al., "Geology and Paleontology of the Upper Miocene Toros-Menalla Hominid Locality, Chad," *Nature* 418 (2002): 152–55.

5. For more information about the Laetoli footprints see http://www.asa3.org/archive/evolution/199505-10/0668.html.

6. From Donald Johanson and James Shreeve, *Lucy's Child: The Discovery of a Human Ancestor* (New York: William Morrow, 1989).

7. Some scientists who have studied Lucy argue that she was a tree-dweller, at least part-time, because she has a bone in her wrist that enabled her to knuckle-walk. The theory is that she would not have retained this bone if she hadn't preserved some of the ambulatory habits of her precursors. But other scientists argue that having the bone doesn't mean it was used. It may simply have been an evolutionary leftover, like wisdom teeth or appendices.

8. For more on this see http://www.webster.edu/~woolflm/maryleakey.html.

9. To read a revealing profile of Mary Leakey visit http://www.sciam.com/article.cfm?articleID=0006E1CC-7860-1C76-9B81809EC588EF21&pageNumber=2&catID=4.

10. Donald C. Johanson and Maitland A. Edey, *Lucy: The Beginnings of Humankind* (London: Penguin, 1981), p. 250.

11. For some disagreement on the humanness of the Laetoli footprints read Ian Tattersall and Jeffrey H. Schwartz, *Extinct Humans* (Boulder, Colo.: Westview Press, 2001).

12. Compared to modern humans, *A. afarensis* was small. Males probably stood about four feet, eleven inches high and weighed about a hundred pounds. Females were even smaller, stood about three feet, five inches tall, and weighed about sixty-two pounds.

13. The fossil record reveals many examples of this from microscopic creatures such as foraminiferans (single-celled protists with shells) to various species of trilobites, even the rapid emergence of a descendant of *Tyrannosaurus rex* known as *Daspletosaurus* (frightful lizard), which roamed the plains of Montana and western Canada during the Cretaceous period, about seventy-five million years ago.

14. Whether it is in insects or human beings, HOX genes are lined up tightly in chromosomes, like pearls on a string. Scientists speculate that this grouping and sequence are necessary for the genes to function in a coordinated way. For example, the HOX gene sitting first in line on the chromosome controls the development of the back part of the brain, the second gene is responsible for the upper part of the neck, and so on along the body axis. If they get out of sequence, the body parts they govern would also be physically expressed out of sequence.

15. See Jeffrey H. Schwartz, *Sudden Origins: Fossils, Genes, and the Emergence of Species* (New York: John Wiley & Sons, 1999).

16. For more information about early forested habitats see http://www.sciencedaily.com/releases/2001/07/010712080455.htm and Kate Wong, "An Ancestor to Call Our Own," *Scientific American* 13(2) (2003), 4–13.

17. The lack of trees also means that the grasslands were short on shade, another powerful reason for standing upright. In the jungle, shade was abundant, and there was little need for hominids to reduce how much of their dark and hairy body they exposed to the sun. But under the hot savanna sun, standing erect shades more of the body from the broiling heat at the same time it exposes more of it to the air. Both have a chilling, well-cooling effect. More on this in chapter 5, "Making Thoughts out of Thin Air."

18. At a speed of two miles an hour, a chimpanzee expends about a third more energy than a human. Over the long haul, that simply wouldn't do. (See Rodman and McHenry, "Bioenergetics and Origins of Hominid Bipedalism," *American Journal of Physical Anthropology* 52 (1980): 103–106.) Later studies supported Rodman and McHenry's argument for the energy efficiency of human bipedal walking as compared to chimpanzee quadrapedal locomotion.

19. Scientists have to be careful when drawing conclusions about an animal's behavior from its anatomy, but the fossil evidence suggests that these robust, big-toothed australopiths relied more on roots and nuts for their nourishment than the gracile apes such as *A. africanus,* which seem better suited to an occasional meal of carcass scraps and mashed bone marrow. We see something similar to this in the eating habits of gorillas

and chimpanzees today. Gorillas are almost exclusively vegetarians, and they have the teeth and jaws to prove it. Chimps, as Jane Goodall's studies have shown, will eat meat when they can, and even cooperatively hunt down bush pigs and small monkeys. *Paranthropus robustus's* vegetarian ways may ultimately have been its undoing, because the line seems to disappear about 1.5 million years ago without, so far, another fossilized peep. The evolutionary advantage apparently fell to apes that could complement their diet of bugs, roots, and berries with animal fat and protein.

Chapter 2: Standing Up—Sex and the Single Hominid

1. Charles Darwin, *The Descent of Man* (Norwalk, Conn.: Heritage Press, 1972), p. 187.
2. See the following for more on this subject:
Berscheid, Ellen, and Harry T. Reis. "Attraction and Close Relationships." In Daniel T. Gilbert, Susan T. Fiske, and Gardner Lindzey, eds., *Handbook of Social Psychology* (New York: McGraw-Hill, 1998), pp. 193–281.
Harper, B. "Beauty, Statute and the Labour Market: A British Cohort Study." *Oxford Bulletin of Economics and Statistics* 62 (December 2000): 773–802.
Fisher, Helen. "Why We Love: The Nature and Chemistry of Romantic Love." New York: Henry Holt, 2004.
Cash, T. F., B. Gillen, and D. S. Burns. "Sexism and 'Beautyism' in Personnel Consultant Decision Making." *Journal of Applied Psychology* 62 (1997): 301–10.
Clark, M. S. and J. Mills. "Interpersonal Attraction in Exchange and Communal Relationships." *Journal of Personality and Social Psychology* 37 (1979): 12–24.
Cunningham, M. R. "What Do Women Want?" *Journal of Personality and Social Psychology* 59 (1990): 61–72.
Singh, D. "Adaptive Significance of Female Physical Attractiveness: Role of Waist-to-Hip Ratio." *Journal of Personality and Social Psychology* 65 (1993): 293–307.
Cunningham, M. R., A. R. Roberts, A. P. Barbee, P. B. Duren, and C. H. Wu. "Their Ideas of Beauty Are, on the Whole, the Same as Ours: Consistency and Variability in the Cross-Cultural Perception of Female Physical Attractiveness." *Journal of Personality and Social Psychology* 68 (1995): 261–79.
De Santis, A., and W. A. Kayson. "Defendants' Characteristics of Attractiveness, Race, and Sex and Sentencing Decisions." *Psychological Reports* 81 (1999): 679–83.
3. Timothy Taylor, *The Prehistory of Sex* (New York: Bantam Books, 1996).
4. S. R. Richards, F. E. Chang, B. Bossetti, W. B. Malarkey, and M. H. Kim, "Serum Carotene Levels in Female Long-Distance Runners, *Fertil Steril.* 43, no. 1 (1985): 79–81; C. H. Wu and G. Mikhail, "Plasma Hormone Profile in an Ovulation." *Fertil Steril.* 31, no. 3 (1979): 258–66.
5. Nikolas, Lloyd, "Why Women Have Breasts" (see http://www.staff.ncl.ac.uk/nikolas .lloyd/evolve/breasts.html).
6. In addition to the work of Rosenberg and Trevathan, C. Owen Lovejoy at Kent State University and Robert G. Tague of Louisiana State University have studied a range of fossil pubic and pelvic bones and concluded that while the birth canals of australopiths are larger than a human's, the baby still would have had to rotate either forward or backward to fit its shoulders through the narrowing canal. This means that sometimes the

baby would have been born facing its mother's back or facing forward. Either way it would have been more difficult than the average chimpanzee's birth, and that would have required some help. From Karen Rosenberg and Wenda R. Trevathan, "The Evolution of Human Birth," *Scientific American* 13(2) (2003). Also see C. Owen Lovejoy, "The Evolution of Human Walking," *Scientific American* (November 1988).

7. Various versions of *Homo habilis* fossils have been found, and many are debated. Some date back as far as 2.3 million years ago.

8. Stephen Jay Gould, *Ontogeny and Phylogeny* (Cambridge, Mass.: Harvard University Press, Belknap Press, 1997), pp. 372–73.

9. Ibid., pp. 352–56, and http://www.serpentfd.org/a/gouldstephenj1977.html for details.

10. Gould, *Ontogeny and Phylogeny,* p. 401.

Chapter 3: Mothers of Invention

1. For a short movie that illustrates these corpuscles visit http://www.microscopyu .com/galleries/confocal/meissnerscorpusclesprimate.html.

2. Shakespeare had an enormous influence on the vocabulary and expression of the English language. He invented more than seventeen hundred words (not to mention countless phrases that we all commonly use every day). He did this mostly by changing nouns into verbs or verbs into adjectives, or by putting two words together that had never before been connected. Sometimes he added prefixes and suffixes, and often he made new words out of whole cloth. A short list of some words the great bard invented include advertising, amazement, arouse, assassination, backing, bandit, bloodstained, bump, buzzer, circumstantial, cold-blooded, compromise, dauntless, dawn, dishearten, drugged, dwindle, frugal, generous, gloomy, gossip, gust, hobnob, impartial, invulnerable, lackluster, laughable, lonely, luggage, lustrous, madcap, majestic, mimic, monumental, moonbeam, obscene, olympian, outbreak, radiance, rant, remorseless, savagery, scuffle, submerge, summit, swagger, torture, tranquil, undress, unreal, worthless, zany. For more visit: http://shakespeare.about.com/library/weekly/aa042400a.htm or read *Coined by Shakespeare* by Jeffrey McQuain and Stanley Mallessone (Springfield, Mass.: Merriam-Webster, 1998).

3. See *In the Shadow of Man* by Jane Goodall for more on the tool use of chimpanzees (and everything else about their lifestyle in the wilds of Africa).

4. John Napier, *Hands,* rev. ed. (Princeton, N.J.: Princeton University Press, 1993), p. 55.

5. Mary Marzke, "Evolution," K. M. B. Bennett and U. Catilello, eds, *Insights into the Reach to Grasp Movement* (Amsterdam: ElsevierScience B.V., 1994), chapter.

6. Kathy D. Schick and Nicholas Toth, *Making Silent Stones Speak: Human Evolution and the Dawn of Technology* (New York: Touchstone Books, 1993).

7. Beginning with excavations on July 21, 1986, in Olduvai Gorge, important portions of *Homo habilis*'s body were found and reconstructed (one key assemblage of fossils is known as OH 62).

Until 1986 *Homo habilis* had been considered ancestral to modern humans and a direct ancestor of *Homo erectus,* so its limb proportions were thought to be similar to modern humans. But the OH 62 skeleton revealed that *H. habilis had* surprisingly ape-like limb proportions.

Modern humans have an upper arm bone (humerus) which is considerably shorter than the upper leg bone (femur). In modern apes the humerus and the femur are nearly the same length. This means that *Homo habilis* had a body structure much more like an ape or *Australopithecus afarensis,* than like a modern human. OH 62 also was small—about three feet tall—and probably a female. This implies a significant amount of sexual dimorphism, again, more like apes or *A. afarensis* than modern humans, where the difference in size between men and women is not so great. Both of these finds were surprising.

8. "We acquire a large system of primary metaphors automatically and unconsciously simply by functioning in the most ordinary of ways in the everyday world from our earliest years. We have no choice in this. Because of the way neural connections are formed during the period of conflation, we all naturally think using hundreds of primary metaphors." *Philosophy in the Flesh* by George Lakoff and Mark Johnson (New York: Perseus, 1998), p. 47.

9. From the work of Srini Narayanan, one of Lakoff's students. For more see http://www.google.com/search?q=Narayanan+neural+theory&ie=UTF-8&oe=UTF-8.

10. P. M. Greenfield, "Language, Tools, and Brain: The Ontogeny and Phylogeny of Hierarchically Organized Sequential Behavior," *Behavioral and Brain Sciences* 14 (1991), 531–51, and P. M. Greenfield. "Language, Tools, and Brain Revisited," *Behavioral and Brain Sciences* (1991): 531–95.

11. We are unique in this among species partly because the mental lives of all animals are shaped by their physical experience. Dolphins and whales sense their surroundings with elaborate sonar echolocation techniques and communicate by rich suites of clicks. In his book *The Dragons of Eden* (New York: Ballantine Books, 1977), pp. 107–8, Carl Sagan pointed out, "One very clever recent suggestion, which is being investigated, is that dolphin/dolphin communication involves a re-creation of the sonar reflection characteristics of the objects being described. In this view a dolphin does not 'say' a single word for shark, but rather transmits a set of clicks corresponding to the audio reflection spectrum it would obtain on irradiating a shark with sound waves. . . . The basic form of dolphin/dolphin communication in this view would be a sort of aural onomatopoeia, a drawing of audio frequency pictures—in this case caricatures of a shark. We could well imagine the extension of such a language from concrete to abstract ideas. . . . It would be possible, then, for dolphins to create extraordinary audio images out of their imaginations rather than their experience." In this sense, echolocation (the dolphin equivalent of hands and eyes) shapes clicks (the dolphin version of language). Whether they have made that leap isn't yet fully understood.

12. Sherman Wilcox, "The Invention and Ritualization of Language" in Barbara J. King, ed., *The Origins of Language* (Santa Fe, N. Mex.: School of American Research Press, 1999).

13. M. A. Arbib, "From Monkey-like Action Recognition to Human Language: An Evolutionary Framework for Neurolinguistics," *Behavioral and Brain Sciences* (revision completed February 1, 2004), with the author's response to the commentaries on the article (completed August 22, 2004).

14. In still another study that Japanese researchers conducted forty-seven functional magnetic resonance imaging (fMRI) that revealed that people adapt when they wear specialized glasses that reversed the images of their right and left hands. When they picked up a ball with their right hand the glasses reversed the image so that it *looked* as though it was actually their left hand doing the grasping. It took participants in the study nearly a month to straighten out the garbled signals their brains were receiving, but again, it turned out that Broca's area was behind most of the straightening out, resyncing the signals being sent from the eyes and the hands. This confirmed that Broca's area is not only central to speech, but also crucial to coordinating how we handle and manipulate objects.

    In another, more recent study, Marco Iacoboni at the Los Angeles School of Medicine found that Broca's area "lit up" when subjects watched another person trying to accomplish a task. "Cortical Mechanisms of Human Imitation," *Science* 286 (1999): 25–26.

15. Charles Darwin, *The Descent of Man, and Selection in Relation to Sex* (Norwalk, Conn.: Heritage Press, 1972).

16. "Meme" is a word coined by Oxford zoologist Richard Dawkins, who has argued that like genes, which are preserved in species because they lead to traits that enable survival, memes are ideas or concepts that survive, flourish, and are adopted in cultures because they work. Riding horses is a meme that was adopted by cultures around the world as a terrifically efficient way to get around. Agriculture is a universally adopted meme, except for those few hunter-gather tribes that persist in remote parts of the world. Watching movies on DVDs or communicating by e-mail also are memes that are flourishing, just as the genes that enabled upright walking, speech, and music are regularly returned and then drawn from the same gene pool.

Chapter 4: Homo hallucinator—the Dream Animal

1. For more on the origins of language see Eric P. Hamp and E. H. Sturtevant, *Linguistic Change: An Introduction to the Historical Study of Language* (Chicago: University of Chicago Press, 1961).

2. From Derek Bickerton, University of Hawaii, in his book *Roots of Language* (Ann Arbor, Mich.: Karoma Publishers, 1981).

3. For a broad and fascinating (but sporadic) survey of nonverbal forms of communication and the science behind them, visit the Web site of the Center for Nonverbal Communication at http://members.aol.com/nonverbal2.

4. Some scientists believe that vocal language may date back only about two hundred thousand years. Even as human primates, we have not fully come to grips with the prolonged, face-to-face closeness required for speech. Speaking to a stranger, for example, stresses our autonomic nervous system's sympathetic (fight-or-flight) system. That then speeds our heartbeat, dilates our pupils, and cools and moistens our hands. The limbic brain's hypothalamus also instructs the pituitary gland to release hormones into the circulatory system, increasing the flow of our blood, sweat, and fears.

    If we're upset, scared, or confused, we may have trouble making eye contact. We affectionately tousle the hair of our children, pat a cheek, and hold hands with a child to protect them or silently stay emotionally in touch with a lover.

5. Daniel McNeill, *The Face* (Boston: Little, Brown, 1998).
6. See http://www.bbc.co.uk/science/humanbody/body/factfiles/facial/frontalis.shtml for more on facial muscles.
7. P. Ekman and W. V. Friesen, "The Repetoire of Non-verbal Behavior: Categories, Origins, Usage, and Coding," *Semiotica* 1 (1969): 49–98. http://face-and-emotion.com/dataface/nsfrept/psychology.html. Basic emotions expressed by the face at http://face-and-emotion.com/dataface/emotion/expression.jsp.
8. See *National Geographic*, May 1997; p. 89.
9. Had Leonardo da Vinci found the skeleton of *Homo erectus* as he wandered the hills of Florence, as he often did back in the fifteenth century, even he, the master of detailed observation, would have had a difficult time realizing that the bones he was examining did not belong to a contemporary. Leonardo was fascinated with human anatomy. You might say he was even obsessed. What remain of his legendary notebooks are filled with drawings of hands and feet and forearms, heads and noses and eyes, each comparing the proportions of the others used as measures to reveal the human body's remarkably symmetrical proportions.
10. While recent evidence indicates that *Homo habilis* made some forays beyond Africa into the Middle East and southern Russia, *H. erectus* roamed even farther. His bones have been found buried in the earth of Indonesia and Australia. Earlier work hinted that improvements in tool technology about 1.4 million years ago—namely, the advent of the Acheulean hand ax—allowed hominids to leave Africa. But new discoveries indicate that *H. erectus* hit the ground running, so to speak. Rutgers University geochronologist Carl Swisher III and his colleagues have shown that the earliest *H. erectus* sites outside of Africa, which are in Indonesia and the Republic of Georgia, date to between 1.8 million and 1.7 million years ago. It seems that the first appearance of *H. erectus* and their initial spread from Africa were almost simultaneous. Why? Food. What an animal eats often dictates how much territory it needs to survive, and carnivorous animals require bigger home ranges than do herbivores of comparable size because they have to roam farther to get the calories they need. (Their food is harder to catch than a plant.)

   Until recently, scientists believed that *Homo erectus* was the first human ancestor to make its way out of Africa, but between 1999 and 2001 paloegeographer Davit Lortkipanidze's team found skull fragments of a creature with a brain about the size of *Homo habilis* in Dmanisi, Georgia (formerly part of the USSR). Though it resembled *H. habilis,* the feeling is this creature falls somewhere between *H. habilis* and *H. erectus* and so was given a new name: *Homo georgicus.*
11. While recent evidence indicates that *Homo habilis* made some forays beyond Africa into the Middle East and southern Russia, *H. erectus* roamed much farther. His bones have been found buried in the earth of Indonesia and Australia.
12. See Rick Gore, "The Dawn of Humans: Expanding Worlds," *National Geographic* 191, no. 5 (1997): 91–92.
13. Ibid., 84–109.
14. For more on *Homo erectus* see http://www.wsu.edu:8001/vwsu/gened/learn-modules/top_longfor/timeline/*Homo erectus*/*Homo erectus*-a.html.

15. In primates the neocortex's corticospinal tract evolved to link the posterior parietal cortex to supplementary motor, premotor, and primary motor cortices to cervical and thoracic anterior-horn spinal interneurons as well as motor neurons that control the arm, hand, and finger muscles for skilled movements such as the precision grip. Just as important, parts of the inferior temporal neocortex evolved to provide visual input that enables us to recognize complex shapes, and the inferior temporal cortex permitting heightened responses to hands and the ability to recognize faces.

  Later evolution of the corticobulbar pathways to the facial nerve enabled intentional facial expressions (such as smiles). Next, scientists believe, Broca's cranial pathways developed Broca's-area neocircuits that ran along corticobulbar pathways to multiple cranial nerves, which resulted in the muscle control that now allows us to speak. It is also possible that Broca's-area neocircuits found their way along the corticospinal pathways to cervical and thoracic spinal nerves that enabled manual sign language and linguisticlike mime cues.

16. Also see D. McNeill, "So You Think Gestures Are Nonverbal?," *Psychological Review* 92, no. 3 (1985): 350–71.

17. Iverson, J.M., O. Capirici, and M.C. Caselli. "From Communication to Language in Two Modalities," *Cognitive Development* 9 (1994): 23–43.

18. See Michael C. Corballis, *From Hand to Mouth: The Origins of Language* (Princeton, N. J.: Princeton University Press, 2003).

19. See Takeshi Nishimura, Akichika Mikami, Juri Suzuki, and Tetsuro Matsuzawa, "Descent of the Larynx in Chimpanzee Infants," *PNAS* 100 (2003): 6930–33.

20. See http://www.abc.net.au/science/news/stories/s862604.htm.

21. Garcia has written a book and produced a videotape that helps parents teach their infants to use sign language before they start talking. See Joseph Garcia, *Sign with Your Baby: How to Communicate with Infants Before They Can Speak* (Bellingham, Wash.: Stratton-Kehl Publications, 2001).

22. Quoted in *New York Times* before her death in 2003.

23. That early hand signaling increases IQ might not seem immediately obvious, but it makes sense. Verbal acuity and IQ are linked. And the parts of the brain that control fine hand motion overlap with parts of the brain that send signals to our lungs, throats, lips, and mouths when we speak. The connection between words and gestures is, neurologically speaking, literal. The studies (Acredelo's study involved 103 children) also revealed that learning and using hand signals helped babies make other transitions to speech. If a hand-signaling child, for example, said "pwease" rather than "please" or "toofbrush" rather than "toothbrush," the studies revealed that they would make the gesture signifying those things until they mastered the word's correct pronunciation. This means, strangely enough, that children who are *not* taught to sign, but who are physically gifted with throats, tongues, and lungs that enable them to speak earlier than most children, may later grow to be more intelligent. In other words, they didn't speak earlier because they were smarter than other children, they became smarter later because they could speak earlier.

24. For online movies that illustrate the different hand movements see http://www.dartmouth.edu/~lpetitto/nature.html.

25. All babies, of course, throw their arms and hands all over the place all the time. So how could the researchers distinguish hand tossing from real hand babbling? They videotaped the infants and used optoelectronic tracking systems to record all of the children's hand movements in three dimensions. These recordings showed that all of the children gestured in rapid, chaotic bursts, but in addition to these, the children whose parents used sign language gestured in very specific ways, more slowly, with their hands placed only in a tightly restricted space in front of their bodies, where all signed language is "spoken."

26. Even when children who speak using ASL manage to get a grip on the complex gestures they use (the equivalent of complex spoken sentences), they continue to struggle with gaining full mastery of the language just as other children struggle with spoken language at the same age. Ursula Bellugi at the Salk Institute found that as late as age ten both speaking and signing children make the same grammatical errors, and in one experiment, both struggled to keep the characters in a story straight when telling that complicated story. Bellugi felt that this was because no matter whether the children are using sign language or their voices to communicate, they still draw on Broca's and Wernicke's areas to process what they are trying to say.

27. Gestures and speech are precisely synchronized (see *Hand to Mouth: The Origins of Language,* by Michael C. Corbalis [Princeton, N.J.: Princeton University Press, 2002]), p. 100. He suggests that speech and gesture form a single integrated communications system, something that also indicates that they share a common neurological mechanism that controls them. Corbalis's take on this is that speech and gesture aren't competing forms of communication, but integrated ones.

   Even when a stroke or a terrible accident eradicates speech completely, a patient often can fall back on gesture and communicate very effectively. But if patients lose the mental capacity to mime or gesture, psychologists generally agree that they have become psychotic or suffer from severe dementia.

28. Another strange twist here. Talking patients who suffer from aphasia can learn to "speak" more effectively using ASL, so the parts of the brain handling signing and speech must not be exactly the same areas, though they clearly draw on very similar parts of the brain. See S. W. Anderson, H. Damasio, A. R. Damasio, et al., "Acquisition of Signs from American Sign Language in Hearing Individuals Following Left Hemisphere Damage and Aphasia," *Neuropsychologia* 30 (1992): 329–40. This shows how adaptable our brains can be. It is as if our stomachs could learn to digest cellulose or tin cans.

29. Pettito and Robert Zatorre (at McGill University) also have studied positron emission tomography (PET) brain scans of eleven profoundly deaf people and ten hearing people. Previous work had already shown that deaf people who communicate using ASL process signed sentences mostly in the left hemisphere of their brain, just as hearing people do when they parse spoken language (both Wernicke and Broca's areas in the brain are almost exclusively in the left hemisphere).

   The PET scans revealed that when most of us who use spoken language are racking our brains to come up with the right word (we all know the feeling), we use a specific structure in the left inferior frontal cortex to capture and express the thought.

What was fascinating was that when the brain scans of both hearing and deaf subjects were compared, they revealed that exactly the same areas of the brain activated when deaf subjects struggled to come up with the right sign! Even when the deaf subjects processed totally meaningless grammatical hand movements, the planum temporale lit up just as it does in hearing people when they try to make sense of random incoming syllables that don't carry the same symbolic meaning that words do.

30. Lawrence Osborne, *New York Times,* October 24, 1999.

31. Ann Senghas, Sotaro Kita, and Aslı Ozyurek, "Children Creating Core Properties of Language: Evidence from an Emerging Sign Language in Nicaragua," *Science* 305 (September 17, 2004).

32. See http://www.dartmouth.edu/~lpetitto/optopic.jpg.

Chapter 5: Making Thoughts Out of Thin Air

1. *Discourse on Method and Mediations,* trans. L. Lafleur (1637) (Indianapolis, Ind.: Bobbs-Merrill, 1960).

2. Another advantage of standing upright is that it reduces the amount of the body you expose to the sun. The long cylinder of a human biped presents a far smaller target to the sun than a gorilla or a lion, which may explain why gorillas live in the rain forest and lions prefer to hunt at night.

3. Marvin Harris, *Our Kind* (New York: Harper & Row, 1990), pp. 52–53.

4. William R. Leonard, "Food for Thought," *Scientific American* 13(2) (2002). By using estimates of hominid body size compiled by Henry M. McHenry of the University of California at Davis, Robertson and Leonard reconstructed the proportion of resting energy needs required to support the brains of human ancestors. They calculated that a typical 80- to 85-pound australopith with a 450-cc brain would have devoted about 11 percent of its resting energy to powering its brain. *H. erectus,* which weighed about 125 to 130 pounds and had a brain about 900 cc in size, would have required 17 percent of its resting energy, or 260 out of 1,500 kilocalories a day.

5. From http://www.anthro.fsu.edu/people/faculty/falk/radpapweb.htm, later published in *The Evolution in Mammals of Nervous Systems,* vol. 5, ed. Todd M. Preuss and Jon H. Kaas (New York: Elsevier-Academic Press, 2004).

6. Physicians Michel Cabanac and Heiner Brinnel figured out this particular problem by massaging a cadaver's skullcap. The blood flowed through the venous network from the outside of the skull to the diploic veins within the cranial bones and then to the inside of the braincase.

7. See http://www.show.scot.nhs.uk/wghcriticalcare/rational%20for%20human%20 selective%20brain%20cooling.htm for an online version of Cabanac and Brinnel's paper.

8. M. A. Baker, "A Brain-Cooling System in Mammals," *Scientific American* 240 (1979): 130–139.

9. Preuss and Kaas, eds., *The Evolution of Primate Nervous Systems.* Also see http://www .anthro.fsu.edu/people/faculty/falk/radpapweb.htm.

10. In other words, could a system that evolved primarily to reduce overheating and therefore removed obstacles to growth also have played a role in feeding the brain so that it could more rapidly add neurons?

11. The human race speaks roughly sixty-eight hundred languages, and whether you were born in a hut in the jungles of Borneo or the Bronx's North Central Hospital, you entered the world capable of uttering every one of them. That includes the clicking sounds that punctuate the language of the !Kung San in the Kalahari Desert, the singsong Mandarin of East China, or the guttural, long, lumber words so common to Germany.

12. Rachel Smith, "Foundations of Speech Communication," October 8 2004; kiri.ling .cam.ac.uk/rachel/8oct04.ppt

13. Terrence W. Deacon, *The Symbolic Species* (New York: W. W. Norton, 1998), pp. 247–50.

14. When we speak, most of the process is still unconscious. We don't contemplate how to make an "s" sound or say the word "the." But we can and do override our normal, visceral breathing patterns when we talk, and obviously speaking is intentional, not unconscious. The easy way we unconsciously talk with one another all the time might be because by the time we reach age seven or eight we are so practiced at it, something like the way a good pianist can sit and play a complex piece of music she knows well without much thought about where and how her fingers hit the keys, it's second nature.

15. It's difficult to get linguists to agree on an exact number because different accents and dialects of English (as well as other languages) blur the line between two different sounds or the same sound being pronounced slightly differently. The meaning associated with those sounds also is important. "In some languages, where the variant sounds of *p* can change meaning, they are classified as separate phonemes—for example, in Thai the aspirated *p* (pronounced with an accompanying puff of air) and unaspirated *p* are distinguished one from the other. "phoneme." *Encyclopædia Britannica,* 2004. Encyclopædia Britannica Premium Service, November 24, 2004, http://www .britannica.com/eb/article?tocId=9059762.

16. The Khoisan of Africa use 141 phonemes, virtually every sound we are capable of making, in their language. Barbara F. Grimes, ed., *Ethnologue: Languages of the World,* 13th ed. (Summer Institute of Linguistics, 1996). Also see http://64.233.161.104 /search?q=cache:Z6Wp6IGHokYJ:salad.cs.swarthmore.edu/sigphon/papers/de-boer97.ps.Z+maximum+number+phonemes+language&hl=en&client=safari.

17. Like gesture and facial expression, prosody has ancient roots. In fact, some aspects of it go all the way back to the gill actions of mouthless Silurian fish that swam in Earth's seas four hundred million years ago.

18. The January 2003 issue of *Neuropsychology,* published by the American Psychological Association, has an interesting article about a study done in Belgium by psychologists interested in how emotions are processed by our minds. At Ghent University, Guy Vingerhoets, Ph.D., Celine Berckmoes, M.S., and Nathalie Stroobant, M.S., knew that the left brain is dominant for language and the right brain is dominant for emotion.

19. "Complementing earlier studies on hand neurons in macaque F5, Ferrari et al. (2003) studied mouth motor neurons in F5 and showed that about one-third of them also discharge when the monkey observes another individual performing mouth actions. The majority of these "mouth mirror neurons" become active during the execution and observation of mouth actions related to ingestive functions such as grasping, sucking,

or breaking food. Another population of mouth mirror neurons also discharges during the execution of ingestive actions, but the most effective visual stimuli in triggering them are communicative mouth gestures (e.g., lip smacking)—one action becomes associated with a whole performance, of which one part involves similar movements. This fits with the hypothesis that neurons learn to associate patterns of neural firing rather than being committed to learn specifically pigeonholed categories of data. Thus a potential mirror neuron is in no way committed to become a mirror neuron in the strict sense, even though it may be more likely to do so than otherwise. The observed communicative actions (with the effective executed action for different "mirror neurons" in parentheses) include lip-smacking (sucking, sucking and lip smacking); lips protrusion (grasping with lips, lips protrusion, lip smacking, grasping, and chewing); tongue protrusion (reaching with tongue); teeth chatter (grasping); and lips/tongue protrusion (grasping with lips and reaching with tongue; grasping). We thus see that the communicative gestures (effective observed actions) are a long way from the sort of vocalizations that occur in speech." From M. A. Arbib, "From Monkey-like Action Recognition to Human Language: An Evolutionary Framework for Neurolinguistics," *Behavioral and Brain Sciences* 28(2) (2005): 105–24.

20. The evolving configuration of our throats was making it possible for our ancestors to create a broader range of sounds than other primates. In fact, without this rearrangement, language as we know it would be out of the question. This is why efforts to teach chimps to speak have failed and why Koko the gorilla uses sign language and symbols when she "talks" rather than her voice.

21. The idea of language being hardwired into the brain as an evolutionary adaptation was first aggressively proposed in the 1950s by linguist Noam Chomsky.

22. From http://www.ling.upenn.edu/courses/Spring_2001/lingoo1/origins.html.

23. See Arbib's work cited above, but also Robin Dunbar, *Grooming, Gossip, and the Evolution of Language* (Cambridge, Mass.: Harvard University Press, 1996), p. 48.

Chapter 6: I Am Me: The Rise of Consciousness

1. More than any other species, we are careworn. We worry. And worrying, in its own way, is extraordinary. We imagine events that have not happened, we dream up in fine detail all of the things that could go wrong. We wonder, we plan, we think the worst, and then try to imagine how we will deal with it. When we worry we are basically trying to predict the future, or even multiple futures. Other animals do not worry. They may feel fear or even anxiety, but they do not worry because they don't have the cerebral capability.

The daughter of worry is planning. And planning is brought to you by the special architecture and chemistry of your frontal lobes, which are by far the largest on the planet, and the most complexly wired. This is the same part of the brain that enables us to plan, imagine, invent, plot, deceive, dissemble, and mourn. None of us could get out of bed, go about our business, or manage the many relationships that require our constant attention without our prefrontal cortex.

Our ability to worry is strangely connected to a phenomenon that psychobiologist Henry Plotkin calls the "uncertain futures problem." Plotkin credits biologist

C. H. Waddington with having noticed that humans live lives that exemplify the uncertain futures problem. But he takes it to new and interesting levels.

To consider that you have an uncertain future, you first have to be able to imagine one. You also have to be freed from the exclusive commands of your genes. The more advanced a brain is (even though it is itself initially a product of your DNA), the more adaptable it is after you are born. The big breakthrough with the human prefrontal cortex is that it is so much more capable of adapting to change than any other brain on Earth, and therefore liberates us from our DNA. This is why we can operate cell phones even though they weren't invented before we were born, or learn to speak English even though we might be Finnish, Indonesian, or Inuit.

2. K. Fleming, T. E. Goldberg, and J. M. Gold, "Applying Working Memory Constructs to Schizophrenic Cognitive Impairment," in A. S. David and J. C. Cutting, eds., *The Neuropsychology of Schizophrenia* (Hillsdale, N.J.: Erlbaum, 1994).

3. Steven W. Anderson, Antonio Bechara, Hanna Damasio, Daniel Tranel, and Antonio R. Damasio, "Impairment of Social and Moral Behavior Related to Early Damage in Human Prefrontal Cortex," *Nature Neuroscience* 2, no. 11 (November 1999): 1032–37.

4. For more on Phineas Gage see http://www.deakin.edu.au/hbs/GAGEPAGE/ Pgstory.htm.

5. The total number of synapses in the cerebral cortex is sixty *trillion;* from G. M. Shepherd, *The Synaptic Organization of the Brain* (New York: Oxford University Press, 1998), p. 6. However, C. Koch lists the total synapses in the cerebral cortex at 240 trillion; *Biophysics of Computation. Information Processing in Single Neurons* (New York: Oxford University Press, 1999), p. 87. For many of the facts and figures in this sidebar also see http://faculty.washington.edu/chudler/facts.html#brain.

6. Christopher Wills, *The Runaway Brain* (New York: HarperCollins, 1993), p. 262.

7. Stephen Pinker, *The Language Instinct* (New York: William Morrow, 1994), p. 368.

8. Oliver Sacks, *The Man Who Mistook His Wife for a Hat* (New York: Touchstone Books, 1998).

9. From Oliver Sacks, *A Leg to Stand On* (New York: HarperCollins, 1984). In some cases patients lose all control of one side of their body, which sometimes seems to take on a mind of its own. In one case a man sometimes found that one of his arms would decide on its own to suddenly start to take his clothes off.

10. G. G. Gallup, "Self-Awareness and the Emergence of Mind in Primates." *American Journal of Primatology* 2 (1982): 237–48.

11. On another level, the immune system also has evolved a way to distinguish between what is self and what isn't. Your immune system has a molecular "understanding" of what is you and what isn't. Anything that isn't recognized as you and enters your body is attacked as an invader. This is why organ transplantation is so difficult, because the donor organ is usually perceived as something that is not "you" and is therefore assaulted. Autoimmune diseases such as arthritis, AIDS, or lupus are examples of occasions when the immune system "misdiagnoses" parts of the body as an outsider and attempts to destroy them, sometimes with lethal results.

12. Most of this information, such as the ongoing activities of our stomachs and livers, or the capillaries in our lower intestines, isn't passed along to the cerebral cortex.

13. Gerald M. Edelman and Giulio Tononi, *Consciousness: How Matter Becomes Imagination* (New York: Penguin Books, 2000), p. 49, and Gerald M. Edelman, *Wider than the Sky: The Phenomenal Gift of Consciousness* (New Haven: Yale University Press, 2004).

14. Edelman also pioneered the now-accepted concept of Darwinian synaptic selection as the driving force in brain development in children and adolescents.

Chapter 7: Words, Grooming, and the Opposite Sex

1. E. B. Keverne, N. D. Martinez, and B. Tuite, "Beta-endorphine Concentrations in Cerebrospinal Fluid of Monkeys Are Influenced by Grooming Relationships," *Psychoneuroendocrinology* 14 (1989): 155–61.

2. Robert M. Seyfarth and Dorothy L. Cheney, "Meaning and Mind in Monkeys," *Scientific American* (December 1992). Nonhuman primates, such as vervet monkeys, seem to communicate in ways that resemble aspects of human speech. But they do not apparently recognize mental states in others. See http://cogweb.ucla.edu/CogSci/Seyfarth.html. Also see Robin Dunbar, *Grooming, Gossip, and the Evolution of Language* (Cambridge, Mass.: Harvard University Press, 1996), p. 68.

3. Jane Goodall, *In the Shadow of Man* (Boston: Houghton Mifflin, 1998).

4. *Machiavellian Intelligence: Social Expertise and the Evolution of Intellect in Monkeys, Apes, and Humans,* ed. Richard W. Byrne and Andrew Whiten, both at the Psychological Laboratory, University of St. Andrews.

5. R. Byrne and A. Whiten, "The Thinking Primate's Guide to Deception," *New Scientist* 116, no. 1589 (1987): 54–57.

6. Dunbar, *Grooming, Gossip,* p. 63.

7. Other studies by Dunbar and his group have shown that we can hold conversations with up to three other people (a total of four). More than that, and they break down. So if three people are talking at a party and two others join in, someone will be left out, or the group will split into two separate conversations.

8. See: http://cogweb.ucla.edu/CogSci/Seyfarth.html for details and references.

9. Chimps spend about 20 percent of their time socially grooming; we spend about 40 percent of our time in social situations, so Dunbar splits the difference and estimates that when we reached a point where we had to devote 30 percent of our time to social interaction, apelike grooming simply wouldn't have worked effectively any longer.

10. Dunbar, *Grooming, Gossip,* pp. 111–14. Dunbar also speculates that the very last of *Homo erectus* may have developed the rudiments of this kind of language, but only as an afterthought.

11. See Steven Pinker, *The Language Instinct* (New York: William Morrow, 1994), p. 314.

12. Dunbar, *Grooming, Gossip,* p. 123.

13. Barbara Strauch, *The Primal Teen* (New York: Doubleday, 2003).

14. Pinker; *The Language Instinct,* p. 369.

15. This is why it is more interesting to read a novel than the phone book or a list of mathematical equations. Stories are about human relationships, and that is what fascinates us. After all, excelling at them is crucial to our happiness and an essential part of everyday life.

16. L. Cosmides and J. Tooby, "Cognitive Adaptations for Social Exchange." In J. H. Barkow, L. Cosmides, and J. Tooby, eds., *The Adapted Mind* (Oxford: Oxford University Press, 1993), pp. 162–228.

17. Try this test. I deal you four cards, two marked with the numbers 8 and 3 and two marked with the letters E and Z. They are all dealt with the numbers and letters showing. Now I tell you that on the other side of the cards you will find different numbers and letters. In fact, the rule is that a card with a vowel on one side will always have an even number on the other side. To prove the rule, which card or cards should you flip over?

    Figuring this out isn't trivial for most people. In fact, about 75 percent of those who take the test get it wrong. Most people either chose the E card or the E card and the 8, even though the rule has nothing to do with what is on the other side of an even-numbered card. The solution is to flip over the cards with E and 3 on them. They reveal all of the possibilities.

    When the problem is presented this way, people are confused because our minds handle social situations better than purely abstract problems. To prove the point, a scientist named Lida Cosmides at the University of California, Santa Barbara, reworked the problem. Instead of dealing cards, she told an experimental group that there are four people sitting at a table. One is sixteen years old and one is twenty; and one is drinking a soft drink and one is drinking a beer. If the legal drinking age is eighteen, with which one of them do you have to check with to see if the law is being broken? The answer was obvious to almost everyone who took the test. Check the sixteen-year-old and the beer drinker. If the sixteen-year-old is drinking beer or the beer drinker is under eighteen, the law has been violated. From Cosmides and Tooby, "Cognitive Adaptations."

18. Theory of Mind is double-edged. It may help us correctly guess what others are thinking, but is also often a source of many of life's misunderstandings because we don't always guess correctly. Othello, for example, (with Iago's help) imagined that Desdemona was cheating on him, and he murdered her for it. But it was all in his imagination. In fact she was devoted and never considered cheating. Xenophobia, racism, terrorism, and war have their roots in the same sorts of misunderstandings.

19. The left hemisphere controls language in 97 percent of right-handers. The right hemisphere controls it in only 19 percent of left-handers. The rest control language from the left side of the brain, or equally in both sides. Pinker, *The Language Instinct,* p. 306.

20. Ibid.

21. See Dunbar, *Grooming, Gossip*, p. 138, for references.

22. Sometimes the left hemisphere's superior ability to handle verbal duties go beyond handling language. Most split-brain patients show word recognition only in the left hemisphere. But a few can use either hemisphere for this task. Even in these cases, the right brain deals with words far less adeptly than the left brain.

    For instance, among those who can process language in either hemisphere, each isolated hemisphere can recognize a specific letter in genuine words more easily than in nonsense words or in random letter strings. But the right hemisphere takes longer than the left to perform this task and requires considerably more time to "make up its mind" as words get longer.

The right hemispheres of split-brain patients also consistently make grammatical mistakes. They struggle with changing verb tenses, constructing plurals, and indicating possessives. Findings such as these support the idea that the left brain harbors an evolved mechanism for understanding grammatical principles common to all spoken languages, according to Gazzaniga.

Some split-brain patients can also verbally identify many items presented to their right hemispheres, which illustrates an extraordinary ability of the split brain to reorganize itself, sometimes resulting in the emergence of limited right-brain speech ten years or more after surgery.

23. Gazzaniga has outlined the evolution of his theories in several fascinating books, including *The Social Brain* (New York: Basic Books, 1984); *Mind Matters* (Boston: Houghton Mifflin, in association with MIT Press and Bradford Books, 1998); and *The Mind's Past* (Berkeley, Calif.: University of California Press, 2000).

24. See http://www.sciencedaily.com/releases/2005/02/050223122209.htm; "The Oldest Homo Sapiens: Fossils Push Human Emergence Back to 195,000 Years Ago," *ScienceDaily* (February 28, 2005). Also see http://www.sciencedaily.com/releases/2005/02/050223142230.htm.

25. This is called the "out of Africa" theory, and it is based on genetic studies of mtDNA or mitochondrial DNA, which is inherited only from women and mutates so predictably that it makes an excellent molecular clock scientists can use to mark the progress of modern humans as they migrated around the planet. There is, of course, debate about this, too. In 2000 a team of Australian scientists headed by Dr. Alan Thorne at the Australian National University studied the DNA of "Mungo Man," whose skeleton was originally found in New South Wales in 1974. Mungo Man lived sixty thousand years ago, and the researchers say his mtDNA does not match the DNA of other humans. He could, they argue, be proof that the "out of Africa" theory is questionable. Perhaps the human race is a potpourri of creatures who evolved in pockets directly from *Homo erectus,* some in Europe, some in Asia and Australia and Africa before they spread out to meet one another and form the modern human race we know today. See http://news.bbc.co.uk/1/hi/sci/tech/1108413.stm.

26. Noam Chomsky. *Syntactic Structures* (The Hague: Mouton, 1957), and *Knowledge of Language: Its Nature, Origin, and Use* (New York: Praeger, 1986).

27. Of course, thinking is not an all-or-nothing proposition. There are many other highly intelligent animals on the planet—whales, dolphins, gorillas, chimps, and orangutans, even squid and crow. All creatures lie along a continuum. Some abstract more clearly than others, some not at all. But none is as cerebrally gifted as we are. The elaborate cultures we have created are the irrefutable evidence.

Chapter 8: Howls, Hoots, and Calls

1. Adams & Kirkevold, "Looking, Smiling, Laughing and Moving in Restaurants: Sex and Age Differences," *Environmental Psychology and Nonverbal Behavior* 3 (1978): 117–21.

2. Dante Alighieri Paradiso (XXVII, 5).

3. According to legend, Douglass recorded much of the original laughter for the machine he invented (also known as Charlie's Box or the Laff Box) from *The Red Skelton Show.*

Because Skelton did so many pantomimes, it was easy for Douglass to record nice, clean snippets of laughter and applause uninterrupted by sounds of the performer.

4. The major part of Freud's work on laughter was published in 1905 under the title *Wit and Its Relation to the Unconscious,* translated by Joyce Crick (New York: Penguin Classics, 2003).

5. Freud felt that if the joke is successful, the source of laughter in the teller and the listener is the same, something he called "the economy of psychic expenditure." At that point the dangerous, subconsciously repressed idea is expressed, but then the light nature of its expression relives the tension and we laugh out of relief, just as a baby laughs out of relief when it turns out that a dangerous situation isn't really as dangerous as it first seemed. Two kinds of pleasure result, he believed: the pleasure of the relief and the plain fun in playing with words in novel and surprising ways.

6. Writer and poet Dorothy Parker was a master of whiplash and one of the great wits of the twentieth century. She never lost an opportunity to show what sort of comedic collisions she could create by juxtaposing two surprising and hilariously unexpected ideas in one sentence. There was the time she reviewed Katharine Hepburn's performance in the 1933 play *The Lake* and wrote: "She delivered a striking performance. It ran the gamut of emotions from A to B." Or her remark when she arrived at a Yale prom and blandly observed, "If all the girls [here] were laid end to end, I wouldn't be at all surprised." Or her advice to the lovelorn: "Don't put all your eggs in one bastard."

7. These findings resulted from the work of Vinod Goel at York University in Toronto and Raymond Dolan of the Institute of Neurology in London. Both men were looking to understand where in the brain the mental shift takes place that creates laughter. To study the problems they used functional MRI to scan fourteen healthy people while they listened to two types of jokes. Half the jokes were "semantic," like the shark joke, the other half were puns. They also told control jokes which were set-ups with punchless punchlines. To their amazement they found that the different jokes were processed in completely different parts of the brain. Semantic jokes used a network in the temporal lobes, but the puns were processed near parts of the brain that handle speech. See "The Functional Anatomy of Humor," *Nature Neuroscience,* vol. 4:3 (March 2001), p 237.

8. The LaughLab computer counted the number of words in every joke that people submitted and found that jokes containing 103 words are the funniest. The winning "hunters" joke is 102 words long.

9. This is the very same part of the of brain that doctors performing prefrontal lobotomies often destroyed to cure a wide variety of mental problems in the 1930s and '40s. Unfortunately, they eventually learned that though lobotomies didn't destroy patients' ability to think or reason, it often robbed them of their personalities and crippled their ability to relate in emotionally subtle ways.

10. Itzhak Fried, Charles L. Wilson, Katherine A. MacDonald, Eric J. Behnke, Division of Neurosurgery and Departments of Neurology and Psychiatry and Biobehavioral Sciences at the UCLA Medical School, "Consciousness and Neurosurgery," *Nature* 391 (February 12, 1991).

11. Daniel N. Stern, *Interpersonal World of the Infant* (New York: Basic Books, 2000).

12. Parents of easily tickled babies may engage in more physical play (since they are positively reinforced by their infants' laughter for doing so). The play that includes tickling then extends to other forms of humorous physical play and eventually to mental play and wordplay, which encourages children to laugh at any kind of humor. A. J. Fridlund and J. M. Loftis, "Between Tickling and Humorous Laughter: Preliminary Support for the Darwinian-Hecker Hypothesis," *Biological Psychology* 30 (1990): 141–50.

13. Darwin, writing in 1872, thought a comfortable social context was important: "the mind must be in a pleasurable condition; a young child, if tickled by a strange man, would scream in fear." Similarly, the writer Arthur Koestler suggested in 1964 that laughter takes place only when the person being tickled views it as a harmless and playful mock attack.

14. R. R. Provine and Y. L. Yong, "Laughter: A Stereotyped Human Vocalization," *Ethology* 89 (1991): 115–24.

15. This is because with each stride the trunk of a four-legged animal has to brace itself with an inward breath when its paw or hoof hits the ground. Otherwise they wouldn't be able to keep the air they need in their lungs.

16. When we run we can breathe as many as four times for every stride, depending on how fast and how long we have been running. Other mammals, however, have no choice but to breathe each time they take a step, which would lead you to believe that when a cheetah is chasing down a gazelle at sixty miles an hour, it is breathing at very high speed. Robert Provine, "Laughter, Tickling, and the Evolution of Speech and Self," *Current Directions in Psychological Science* 13(6) (December 2004): 215.

17. One indication that our new anatomy shapes the nature of our laughter is that usually the muscles that control speech take ascendancy when we consciously decide they should. Usually we can stop laughing, gain control over our breathing, and speak when we want. But there are times when we laugh so hard that the primal mechanics refuse to give up control and we just can't get a word out until the laughter subsides.

18. At least part of this piece is an extension of Morris's early theory that laughter evolved from "play" attacks, scary situations that are, in fact, not scary. Smiling and laughter are ritualized versions of attack, so these expressions are related to play in animals where facial expressions and reactions are ritualized. The laughter comes from the relief in the tension when an animal figures out the bad situation isn't real. In our case, having expressive faces helps send even clearer messages.

19. This is also true of crying and sobbing. It is often difficult to talk and cry at the same time. That we can't control laughing and crying means that the motor systems that command lips and tongue, diaphragm and lungs are not under conscious control of the parts of the brain that generate the sounds of laughter. Speech inverts this relationship. It controls capabilities such as breathing and prosody and modifies vocal output in extremely subtle and facile ways. In other words, though laughter and crying as we know them may have evolved after speech, or along with it, thei roots go back to much earlier forms of nonconscious communication, such as hoots and calls.

20. See Jane Goodall's book *The Chimpanzees of Gombe: Patterns of Behavior* (Cambridge, Mass.: Harvard University Press, 1986).

21. N. Cousins, *The Anatomy of an Illness* (New York: Bantam Doubleday Dell, 1991), and "The Laughter Connection" in *Head First: The Biology of Hope and the Healing Power of the Human Spirit* (New York: Penguin Books, 1989).

22. "Psychoneuroimmunology of Laughter." An interview with Lee Berk, Dr. PH, from *Journal of Nursing Jocularity* 7, no. 3 (1997): 46–47; http://www.jesthealth.com/art26jnj9.html.

23. Robert R. Provine, "Laughter," *American Scientist* 84, no. 1 (1996): 38–47.

24. According to *Esquire* magazine (February 7, 1999), more than anything else, women want men to make them laugh.

25. See "Reconsidering the Evolution of Nonlinguistic Communication: The Case of Laughter" by Michael J. Owren, Department of Psychology, Cornell University, and Jo-Anne Bachorowski, Department of Psychology, Vanderbilt University. Their basic theory is that laughter evolved from animal calls, but they don't believe as others do that animals call out with a specific message in mind. They theorize that animals call out to influence fellow animals to do a certain thing. So a gorilla may scream before he hits and then learns eventually that the scream will be as good as hitting. But he doesn't initially scream to scare those around him. In other words, the call doesn't have a specific, symbolic meaning. It's just there to evoke a reaction. This, they believe, also applies to laughter.

Chapter 9: The Creature That Weeps

1. According to at least one study, most of us entered the world wailing at C or C-sharp, the sound most easily heard by the human ear and the key at the center of the piano. See Tom Lutz, *Crying, A Natural and Cultural History of Tears* (New York: W. W. Norton, 2001), p.161.

2. S. Chevalier-Skolnikoff, "Facial Expression of Emotion in Nonhuman Primates," in P. Ekman, ed., *Darwin and Facial Expressions* (New York: Academic Press, 1973), pp. 11–89.

3. This is also why crying is so difficult to fake. We can lie with sincerity, but we have a hard time prevaricating when it comes to tears. Even actors who work to cry on cue generally have to call up some feeling that taps deep emotions to bring tears on. We don't naturally have control over crying. Related to this is Terrence Deacon's thought from his book *The Symbolic Species* (New York: W. W. Norton, 1998), p. 236. "The reflex like links between perceiving and producing calls, and the emotional states associated with them, are made evident by the 'infectiousness' of some of our own species' innate calls, specifically laughter and crying."

4. According to a survey conducted by Rupert Sheldrake; see http://www.sheldrake.org/papers/Telepathy/babies.html for more.

5. See http://www.24hourscholar.com/p/articles/mi_m1175/is_n1_v30/ai_19013604#continue.

6. Zahavi is an Israeli biologist whose idea was ridiculed when he first put it forward in 1975, but he has recently been vindicated by some clever mathematical modeling by Alan Grafen at Oxford University. Zahavi and Grafen state that in any encounter in animals where advertisement is important—and that's very, very often—an advertisement is believed only if it's validated by being costly.

7. For more insights into "crying wolf," check the work of Dario Maestripieri, "Parent-Offspring Conflict in Primates," *International Journal of Primatology* 23, no. 4, August 2002.

8. For more on babies and crying, please see http://www.signonsandiego.com/-uniontrib/20050316/news_1c16crying.html.

Chapter 10: The Language of Lips

1. During the Renaissance Italian upper-class women made themselves more attractive by taking belladonna (which means beautiful woman) to dilate their pupils. Unfortunately, belladonna is also poisonous, so in the short term it may have had the desired effect, but in the long run it may not have been such a good idea.

2. According to one Web site (http://www.coolnurse.com/kissing.htm), Dr. Peter Gorden, dental adviser at the British Dental Association, says, "After eating, your mouth is full of sugar solution and acidic saliva, which cause plaque build up. Kissing is nature's own cleaning process. It stimulates saliva flow and brings plaque levels down to normal."

3. This according to Professor Gus McGrouther, who is head of plastic reconstructive surgery at University College, London. McGrouther studied the mechanics of kissing to help him find a solution to overcoming oral deformities in patients, and his research has helped sufferers of Bell's palsy.

4. Remember something similar may be true of women's breasts, too. They may be frontal recapitulations of females' rumps. See chapter 1.

5. Meg Cohen Ragas and Karen Kozlowski, *Read My Lips: A Cultural History of Lipstick* (San Francisco: Chronicle Books, 1998).

6. In an essay titled "Loathsomeness of Long Haire," published in 1653.

7. Kristoffer Nyrop, *The Kiss and Its History* (Auburn, Calif.: Singing Tree Press, 1968); Diane Ackerman, *A Natural History of the Senses* (New York Vintage Books, 1990).

8. Women who were taking birth control pills and were, therefore, basically infertile, did not prefer shirts belonging to men with complementary immune systems. They preferred men whose systems were similar to theirs. The women in the test also said that the T-shirts they liked most often reminded them of former boyfriends. Is there a pattern here?

9. See http://www.antecint.co.uk/main/rm/boarmate.ram.

10. See:http://www.mum.org/mensy71a.htm; Matha K. McClintock, "Menstrual Synchrony and Suppression," *Nature* 229 (1971): 244–245.

11. According to Assistant Professor Jianzhi Zhang from the University of Michigan, the evolution of color vision eliminated any need for pheromones to attract mates. In a 2003 paper published in *Proceedings of the National Academy of Sciences* (June 17, 2003), Zhang argued that color vision may have enabled male monkeys as well as our early ancestors in Africa and Asia to notice subtle changes in female sexual skins. Zhang's team concluded that though humans and some apes still carry genes that should create pheromone receptors in our noses, they have mutated and no longer function.

12. Randolph E. Schmid, "Gay Men Respond Differently to Pheromones," Associated Press, May 10, 2005.

13. The hypothalamus is a hub that links the nervous system to the endocrine system, and it is a great illustration of the entangled and interwoven nature of the brain and its relationship to our bodies. It manages to communicate with our bodies by synthesizing and secreting neurohormones, sometimes called releasing hormones, that stimulate the secretion of still other hormones from the anterior pituitary gland. One of these is known as gonadotropin-releasing hormone (GnRH). Neurons that secrete GnRH are linked to the limbic system, which is deeply involved in the control of both sex and emotions.

14. See Nicholas J. Perella, *The Kiss: Sacred and Profane* (Berkeley: University of California Press, 1969).

15. Raj Kaushik, "Science of a Kiss," *Toronto Star,* February 10, 2004.

16. Richard J. Haier, Rex E. Jung, Ronald A. Yeo, Kevin Head, and Michael T. Alkire, "The Neuroanatomy of General Intelligence: Sex Matters," *NeuroImage* 25 (2005): 320–27.

17. Doreen Kimura, "Sex Differences in the Brain," *Scientific American* 12(1) (2002): 32–37.

18. Simon Baron-Cohen, *The Essential Difference: The Truth About the Male and Female Brain* (New York: Perseus, 2003).

19. There is no direct evidence, but it is interesting that this same area of the brain is so near the area where the first mirror neurons evolved.

20. Kimura, "Sex Differences in the Brain."

21. On the other hand, these very same traits could shape a mind so thoughtful that it might tend to fret too much about relationships, and overly dwell on what could go wrong with them. This might offer some clue as to why women are, as a group, more prone to depression than men.

22. Aeons ago it might have helped alert them to the possibility of abandonment while they were busy raising the children. Today, however, there's a clear downside. Ruminators are unpleasant to be around, with their oversize need for reassurance. Of course, men have their own ways of inadvertently fending off people. As pronounced as the female tilt to depression is the male excess of alcoholism, drug abuse, and antisocial behavior.

    University of Michigan psychologist Susan Nolen-Hoeksema, Ph.D., has found that women ruminate over upsetting situations, going over and over negative thoughts and feelings, especially if they have to do with relationships. Too often they get caught in downward spirals of hopelessness and despair.

23. In what was called the "trial of the century," Thaw's defense immediately made "brainstorm" one of the English language's newest words.

24. Christine R. Harris, "The Evolution of Jealousy," *American Scientist* (January–February 2004): 61–71.

25. The methods for doing in a rival sibling are inexhaustible, even for toddlers. When Hannah, my youngest daughter, was born, her three-year-old sister Molly had an idea as we headed off to the hospital to pick up her mom and the new sibling. "Let's throw Hannah down the steps," she said. When I explained why that might not be a good idea, she thought about it a second and then said, "Okay, only halfway down the steps."

26. Harris, "The Evolution of Jealousy," 61–71.

27. In their book *A General Theory of Love* (New York: Vintage, 2001).

Index

NOTE: An "*n*" after the page number indicates an endnote.

ASPM (Abnormal Spindle-Like Microcephaly Associated) gene, 125
australopithecines
  childbearing, 33–39, 39–40, 226n6
  family tree, 6–7, 10–13
  gait of, 17
  learning process, 122–23
  skeletal proportions of, 227–28n7
  throats of, 90–91
  as tool users, not toolmakers, 48–49, 53
  vegetarian diet of, 225n19
  See also savanna apes; entries beginning with Australopithecus
Australopithecus aethiopicus, 11
Australopithecus afarensis (Ethiopia and Tanzania)
  about, 6, 10–11, 225n12
  Laetoli footprints, 15–17
  Lucy, 13–14, 17, 22, 224n7
  skeletal proportions of, 22–23, 227–28n7
  See also savanna apes
Australopithecus africanus (South Africa), 7, 10–11, 32–33, 37–38. See also savanna apes
Australopithecus anamensis (Kenya), 6, 10
Australopithecus bahreighazali (Chad), 6
Australopithecus boisei, 11
Australopithecus garhi (Ethiopia), 7, 10
Australopithecus robustus, 11, 88
autistics, 161, 194
autoimmune diseases, 236n11
autonomic nervous system, 173, 229n4
axons and dendrites, 109–10

babies' development
  crying and laughter, 149–50, 166, 169, 177, 242n1
  of gestures, 74, 75–78, 231n24, 231n26, 232n27–28
  of jealousy, 202
  neoteny, 33–37
  tickling, play, and humor, 240n12
  toddlers' crying, 177–78
baboons, 30, 126
Bachorowski, Jo-Anne, 160–61, 241n25
Baron-Cohen, Simon, 194
basal ganglia, 97–98, 116
basal tears, 168
Bates, Elizabeth, 76
B cells, disease-fighting, 158
Bell, Charles, 47
belladonna, 242n1(ch. 10)
Bellugi, Ursula, 232n27
Bergman, Ingrid, 185
Bergman's rule, 86
Bible, the, 158, 192
Bickerton, Derek, 64–65, 137

"big bang" theory of language, 137
big toe. See hallux magnus
body language, 66–67, 179
Bolk, Louis, 36–37
Bow-Wow theory of language, 63–64
brain/brains
  about, 107–10
  adaptability of, 232n29, 235–36n1
  aphasia of, 78, 127, 232n28–29
  cooling system, 87–89, 233n6, 233n10
  coordination required for speech, 96, 97–99, 127, 134–35, 238n22
  development in babies, 149–50
  differentiation of left and right hemispheres, 132–35, 148, 238n19, 238n22
  dopamine, 97, 106, 156
  energy consumption of, 87, 129–30, 233n2
  evolution in 3-dimensional space, 54–55, 57–58, 228n8
  facial expressions and, 70, 92, 97
  and feelings leading to tears, 169–72
  genes for wiring the, 223n2
  growth potential during evolution, 17, 36, 39, 49, 58, 125, 136
  intellect, 47–48, 53–55, 57, 138–39, 179, 196–99
  limbic system, 196–99, 201–5, 243n13
  memory, 47–48, 56, 104–7, 136–38, 197
  neoteny as solution to size of, 33–37
  neuroanatomy, 61, 115–18, 123, 193–96
  paleocircuits in, 67
  plasticity of, 35
  processing responses to jokes, 146–47, 240n7, 241n19
  split-brain patients, 133–35, 148, 238n22
  and VNOs, 189
  Wernicke's area, 61, 77–78, 126–27, 146, 232n27
  See also Broca's area; memory; specific parts of the brain
brain damage
  aphasia, 78, 127, 232n28–29
  lobotomies, 105–6, 240n9
  and PLC, 148–49
  to posterior portions of right hemisphere, 113
  to prefrontal cortex, 105
  stroke victims, 78, 113, 232n28–29
brain scans
  of Broca's and Wernicke's areas, 61, 77–78, 95–96, 127, 232n30
  fMRI, 95–96, 229n14, 234n19, 240n7
  of men's and women's brain anatomy, 193–96
  MRI, 147
  of people laughing, 146–47, 240n7

PET scans, 95–96, 232n30, 234n19
    of prefrontal cortex, 104–5
    See also neuroscience
brain size
    of apes, 36, 38
    of australopithecines, 10–11
    at birth, 36
    and cranial air conditioner, 88–89
    and energy consumption, 233n2
    of Homo erectus, 12, 72, 73, 77, 87–88
    of Homo floresiensis, 12
    of Homo habilis, 12, 34–35, 49
    relationship complexity and, 89, 124–25
    troop size and, 123, 125–26
breast-feeding mothers, 169
breasts, female, human versus primate, 28,
    29–30
Brinnel, Heiner, 233n6
British Columbia (Burgess Shale), 19–20
Broca, Pierre Paul, 61
Broca's aphasia, 61
Broca's area
    about, 61, 98, 230–31n15
    in Homo erectus, 73, 98, 99
    in Homo habilis, 132
    learning syntax and, 232n27
    manipulation of objects, 60, 229n14
    muscle control for speech from,
        230–31n15
    proto-Broca's area, 98–99, 132
    theories about speech and, 64–65
    Wernicke's area and, 77–78, 126–27
Bronowski, Jacob, 36
Burgess Shale (British Columbia), 19–20
bushmen of the Kalahari, 24
buttocks, human versus primate, 28
Byrne, Richard, 122

Cabanac, Michel, 233n6
CB (landscaper from IOWA), 148
cerebellum, 116
cerebral cortex
    about, 107
    areas controlled by, 92
    ASPM genes and, 125
    endorphins and, 156
    prefrontal cortex and, 203
    selected information to, 117
    synapse activity in, 236n5
    thalamus and, 115–16
Chad, Africa, 6, 9, 223n3(ch. 3)
Charlie's Box, 239n3
chemical makeup of tears, 168–69
chemical reasons for crying, 170
chemical structure of the brain, 108–10
Cheney, Dorothy, 120, 126, 237n2

childbearing, 33–39, 39–40, 192–93, 226n6
children
    deaf-mute, creating language, 79–82,
        100–101
    infants' development of gestures, 74, 75–78,
        231n24, 231n26, 232n27–28
    infants' development of speech and syntax,
        99–100
    puzzle solving by, 55–57
chimpanzees
    about, x, 3–4, 8
    facial expressions, 153–54
    feeding young, 191
    grooming each other, 119–20, 125, 237n9
    human and chimp foot compared, 18–19
    inability to knowingly point, 74
    involuntary food call, 155
    as meat eaters, 225–26n19
    opposable thumbs, 46
    pelvic saddle and spine, 22
    physical limits on speech, 72–73
    playing and panting, 152–53, 241n15
    polygenous mating system, 40
    socializing, 118–19, 237n9
    and tears, 178
    See also human beings; entries beginning with
        Homo
Chomsky, Noam, 137, 235n21
clusters of neurons
    about, 116, 135–36
    and "big bang" theory of language, 137
    control of unconscious activities, 92
    injuries to, 113
    for laughter and humor, 146–47
    for reading intentions of others, 194–95
color vision, 243n11
communication
    babies' wailing vocabulary, 166, 177
    of "counterproductive" behaviors, 174–75
    with eloquence, 130
    with faces, 67–71, 230–31n15
    future possibilities, 210
    miming, 73–74, 79
    mirror neurons and, 60, 61–62
    mouth gestures, 234–35n19
    nonverbal, 58–62, 65–67, 229n4
    with pheromones, 189–91
    of primates, 4
    tears as, 168, 172, 178–79
    See also crying; facial expressions; gestures;
        kissing; language
communication system, speech and gesture as
    integrated, 232n28
compatibility and laughter, 160–61
Complement 3, 158
conflation of experiences, 55

body's cooling ability, 86
as digital species, 211–12
experience modification, 203
family tree, 6–7, 8–9, 9–13
as fetal primates, 35, 36–38
as freaks of nature, ix
outpacing evolution, 197–98
and pheromones, 186–90, 243n8, 243n11
primal drives and limbic system, 196–99, 205
unique traits of, xi–xiii
See also chimpanzees; savanna apes; *entries beginning with* Homo
human genome, 223n2
See also genetics
humor. *See* laughter
hypothalamus, 191, 243n13

Iacoboni, Marco, 229n14
imagination, 235–36n1
immune system, 187–88, 236n11, 243n8
immunoglobulin antibodies, 158
indexical memory, 106–7
Indonesia *(Homo floresiensis)*, 7, 12
"the Indonesian valve," 4–5, 8
infants. *See* babies' development
inferior temporal neocortex, 230n15
inhibitions, 104–6
insects and pheromones, 187
intangible concepts in concrete terms, 54–55, 57–58, 228n8
intellect, 47–48, 53–55, 57, 138–39, 179, 196–99
interpersonal feedback loop, 130, 237n15
interpreter of the brain, 135–36, 138–39
ions, 108–9
IQ development, early ASL and, 76, 231n24

Japanese research, 229n14
jealousy and revenge, 198–99, 201–2
Johanson, Donald, 13–14, 17
Johnson, Mark, 54–55, 57–58, 228n8
joke about hunters, 145
jokes versus puns, processing of, 146–47, 240n7

K. rudolfensis (Eastern Africa), 7
Kalahari Desert, South Africa, 31, 32–33
Kegl, Judy, 79–80
Kenya, Africa, 6, 10, 71, 73, 223–24n3
Khoisan people (Africa), 234n16
kinesics, 66
kissing
    about, xi–xii
    benefits of, 184, 242n1–2
    bonding aspect of, 204

as collision of lust and love, 197
as cultural conditioning, 185–86
evolution of, 191
feeding practice theory, 191–92
lips for, 183–85
risks of, 192
knowledge-pooling, 61–62
Koestler, Arthur, 240n13
Krogman, W. H., 39
Kurzweil, Ray, 211, 223n2

labia and lips, 184–85
lacrimal gland, 166, 168–69, 177, 178, 179
Laetoli footprints (Tanzania), 15–17
Laff Box, 239n3
Lahn, Bruce, 125
Lakoff, George, 54–55, 57–58
landscaper from IOWA (CB), 148
language
    dolphin/dolphin communication, 228n11
    emotional aspects of, 128–30
    evolution of, 79–82, 100–101, 132, 137–39, 178–79
    inadequacy of, 205
    intangible concepts into, 54–55, 57–58, 228n8
    metaphors, 54–55, 65, 81, 228n8
    organizing symbols into, 106–7, 110–11, 126, 128
    origin theories, 63–65, 173
    phonemes, 93–94, 96, 234n15–16
    prosody, 96–97
    recursion, 110–11
    sign language, 75–78, 79–80, 231n24, 231n26, 232n27–28, 232n30
    for social issues, 98–99, 125–26, 129–31, 237n17
    women's facility with, 193–94
    See also Broca's area; communication; gestures; speech; syntax
Language Instinct, The (Pinker), 132, 238n19
Lannon, Richard, 202
larynx, 72, 90, 92–93
larynx of infants, 75, 99
LaughLab project, 145–47, 240n8
laughter
    about, xi–xii, 143–44
    baby's crying becomes, 149–50
    chimps form of, 152–53
    Freud on, 145, 239n5
    healing potential of, 156–58
    muscle control during, 241n17
    pathological laughter and crying, 148–49
    play and, 150–52
    premeditated, 146
    social nature of, 158–61

theory of punctuated equilibrium, 19–21, 225n13

thinking, 106–7, 188, 196, 244n22. *See also* prefrontal cortex

Thorne, Alan, 239n25

thoughts, verbalizing, 98

three-dimensional space and evolution, 54–55, 57–58, 228n8, 228n11

"three-jawed chuck" grip, 50

throat. *See* pharynx and throat

thumbs
    about, xi–xii, 45–47
    mind adaptations related to, 47–48, 53, 62
    and toolmaking, 48–51

tickle reflex, 150–52, 240n12

Tobias, Philip, 49

toddlers' crying, 177–78

toes, xi–xii, 4, 16, 18–19, 25. *See also* hallux magnus

ToM (Theory of Mind), 131, 160, 238n18

tool makers
    brain's adaption to, 58
    gestures, speech, and, 61, 95, 232n28
    hand ax, 71–72
    nature's lack of distinguishing, 210–11
    in Olduvai Gorge, 49, 51, 227n7(ch. 3)
    stone tools, 51–52
    tools making the makers, 209–10
    tool users versus, 4, 48–49, 53

Toth, Nicholas, 51–52

Trevathan, Wenda, 33–34

"the troop within our heads," 131

trust, 130, 161, 174–75

T-shirt research, 186–88

Turkana Boy (Kenya), 71, 73

20 / 20 (TV program), 190

UCLA School of Medicine, 147–48, 157–58

ulnar opposition, 46–47

uncertain futures problem, 235n1

University of California at Irvine, 193

University of New Mexico, 193

University of Rochester School of Medicine, 147

Up from Dragons (Skoyles and Sagan), 130–31

ventromedial prefrontal cortex, 147

vervet monkeys, 120, 126

visual cortex, 117

visualization as mnemonic device, 53–55

VNOs (vomeronasal organs), 187, 189, 191

vocal folds/cords, 93, 96

vocal lekking, 129, 160

vomeronasal organs (VNOs), 187, 189, 191

Waddington, C. H., 235n1

Walcott, Charles Doolittle, 19

Walker, Alan, 71, 73

walking. *See* hallux magnus; standing upright

Wedekind, Claus, 186–88

Wernicke, Karl, 127

Wernicke's area, 61, 77–78, 126–27, 146, 232n27

What's Bred in the Bone (Davies), 47

whining versus crying, 177–78

whiplash effect, 145–46, 150, 155, 240n6, 241n19

Whiten, Andrew, 122

Wilson, Frank, 74

Wiseman, Richard, 144–47, 240n8

women
    brain-driven focus of, 193–94
    and crying, 169, 170
    evaluating men, 129, 200–201
    as hardwired for socializing, 244n20, 244n22
    and laughter, 159–61
    lips and labia, 184–85
    men's evaluation of, 28–30, 196, 199–201
    menstrual cycle and pheromones, 189, 190
    and pregnancy, 33–35, 195
    and socializing, 194–95

working memory, 56, 104–6

worrying, 103, 235n1

Zahavi, Amotz, 174–75, 242n6

Zhang, Jianzhi, 243n11

About the Author

Chip Walter is a science journalist, documentary filmmaker, and former CNN bureau chief. He is coauthor (with William Shatner) of *I'm Working on That* and author of *Space Age*. Walter teaches science writing at Carnegie Mellon University, is senior manager of strategic communications at the University of Pittsburgh Medical Center (UPMC), and lives in Pittsburgh, Pennsylvania, with his two daughters, Molly and Hannah.